CAMBRIDGE MONOGRAPHS ON APPLIED AND COMPUTATIONAL MATHEMATICS

42 **High-Accuracy Finite Difference Methods**

The *Cambridge Monographs on Applied and Computational Mathematics* series reflects the crucial role of mathematical and computational techniques in contemporary science. The series publishes expositions on all aspects of applicable and numerical mathematics, with an emphasis on new developments in this fast-moving area of research.

State-of-the-art methods and algorithms as well as modern mathematical descriptions of physical and mechanical ideas are presented in a manner suited to graduate research students and professionals alike. Sound pedagogical presentation is a prerequisite. It is intended that books in the series will serve to inform a new generation of researchers.

A complete list of books in the series can be found at www.cambridge.org/mathematics.

Recent titles include the following:

26. A practical guide to the invariant calculus, *Elizabeth Louise Mansfield*
27. Difference equations by differential equation methods, *Peter E. Hydon*
28. Multiscale methods for Fredholm integral equations, *Zhongying Chen, Charles A. Micchelli & Yuesheng Xu*
29. Partial differential equation methods for image inpainting, *Carola-Bibiane Schönlieb*
30. Volterra integral equations, *Hermann Brunner*
31. Symmetry, phase modulation and nonlinear waves, *Thomas J. Bridges*
32. Multivariate approximation, *V. Temlyakov*
33. Mathematical modelling of the human cardiovascular system, *Alfio Quarteroni, Luca Dede', Andrea Manzoni & Christian Vergara*
34. Numerical bifurcation analysis of maps, *Yuri A. Kuznetsov & Hil G.E. Meijer*
35. Operator-adapted wavelets, fast solvers, and numerical homogenization, *Houman Owhadi & Clint Scovel*
36. Spaces of measures and their applications to structured population models, *Christian Düll, Piotr Gwiazda, Anna Marciniak-Czochra & Jakub Skrzeczkowski*
37. Quasi-interpolation, *Martin Buhmann & Janin Jäger*
38. The Christoffel–Darboux kernel for data analysis, *Jean Bernard Lasserre, Edouard Pauwels & Mihai Putinar*
39. Geometry of the phase retrieval problem, *Alexander H. Barnett, Charles L. Epstein, Leslie Greengard & Jeremy Magland*
40. Discrete variational problems with interfaces, *Roberto Alicandro, Andrea Braides, Marco Cicalese & Margherita Solci*
41. Spectral and spectral element methods for fractional ordinary and partial differential equations, *Mohsen Zayernouri, Li-Lian Wang, Jie Shen & George Em Karniadakis*

High-Accuracy Finite Difference Methods

BENGT FORNBERG
University of Colorado Boulder

CAMBRIDGE
UNIVERSITY PRESS

CAMBRIDGE
UNIVERSITY PRESS

Shaftesbury Road, Cambridge CB2 8EA, United Kingdom

One Liberty Plaza, 20th Floor, New York, NY 10006, USA

477 Williamstown Road, Port Melbourne, VIC 3207, Australia

314–321, 3rd Floor, Plot 3, Splendor Forum, Jasola District Centre, New Delhi – 110025, India

103 Penang Road, #05–06/07, Visioncrest Commercial, Singapore 238467

Cambridge University Press is part of Cambridge University Press & Assessment, a department of the University of Cambridge.

We share the University's mission to contribute to society through the pursuit of education, learning and research at the highest international levels of excellence.

www.cambridge.org
Information on this title: www.cambridge.org/9781009566537

DOI: 10.1017/9781009566544

When citing this work, please include a reference to the DOI 10.1017/9781009566544

First published 2025

Cover image: Adrean Webb, Tokyo Institute of Technology

A catalogue record for this publication is available from the British Library

A Cataloging-in-Publication data record for this book is available from the Library of Congress

ISBN 978-1-009-56653-7 Hardback

Contents

Preface

What This Book Is About

A finite difference (FD) formula is a local approximation, typically of a derivative, and it consists of weights that can then be applied to the values of arbitrary functions. Methods based on FD approximations are ubiquitous in modern scientific computing. They were well established for solving differential equations over a century ago, when the word "computer" referred to a person with plenty of paper, pencils, and persistence. From at first amounting to a quite straightforward use of simple formulas, readily obtained from a few terms in a Taylor expansion, many specialized FD procedures and refinements have since been developed. Finite element methods evolved around 1960, and software packages based on these are nowadays widely used, especially for many engineering applications. Other offsprings include finite volume, spectral element, and discontinuous Galerkin methods. Two additional major developments began in the early 1970s. One was to push FD stencils toward increasing sizes and orders of accuracy, leading to pseudospectral (PS) methods. The other, focusing on geometric flexibility and allowing for grid-free discretizations, replaced in their derivations polynomials with radial basis functions (RBFs). From the latter have more recently evolved RBF-generated FD methods (or RBF-FD for short), offering FD-like usage, high orders of accuracy, together with total geometric flexibility and easy implementation in any number of space dimensions.

There exists already an extensive literature in the forementioned areas. However, one theme that has not yet received as much attention is the rich middle ground between low-order and extremely high-order FD-based approximation methods. Since traditional FD methods (still a workhorse for general scientific computing) are closely related also to approximations of integrals and of fractional derivatives, some such applications are also described. This book is additionally motivated by several recent developments in these latter areas. In spite of its long history, the FD topic is still a rapidly evolving one.

What This Book Is Not About

The methods described here are centered around spatially local approximations to differential operators and include also some additional contexts where FD-inspired tools are utilized. Nodes used for discretizations may be grid based or irregularly distributed – but if so only when that is motivated by physical reasons (such as irregularly shaped boundaries or a need for locally higher resolution) rather than as artifacts of any special numerical method. In particular, we do not discuss any methods related to orthogonal polynomials, where boundary node clustering is employed purely to control the polynomial Runge phenomenon. FD spin-offs (such as finite elements, finite volume, and discontinuous Galerkin) and applications to integral equations are also not described. Furthermore, this book is not a place to look for functional analysis or for elaborate error estimates.

Brief Summary of the Main Chapters

1. **Introduction to FD Methods:** The history of FD approximations goes back further than that of calculus. The classical definition of a derivative is in itself an example of a very simple (and quite inaccurate) FD formula. While many of its basic properties follow quite immediately from Taylor expansions, numerous additional perspectives are very helpful in appreciating the method's strengths (and weaknesses).
2. **Brief Summary of PS Methods:** FD approximations of increasing orders of accuracy require larger stencil sizes. The limiting case has numerous important applications, which have been extensively treated in the literature. Our present summary aims more to provide some general insights into the pros and cons of pushing up the accuracy order than to describe PS implementations and technicalities.
3. **FD Approximations for Ordinary Differential Equations (ODEs):** The most common reason for wanting to approximate derivatives is to apply these to the solution of differential equations. In the case of ODEs, many of the well-established procedures are immediately related to

FD approximations – often more closely than may be apparent from how these are customarily described.

4. **Grid-based FD Approximations for Partial Differential Equations:** Although a variety of PDE solvers have been developed for different applications, quite straightforward FD-based approximations remain of great utility and importance. As for ODEs, high orders of accuracy often significantly increase computational efficiency.
5. **Mesh-Free FD Approximations:** In all the cases above, the core concept behind the FD approximations has been Taylor expansions. These are easy to work with and are in some sense optimal in their representations of functions locally around a single point. However, polynomial-based approximations in more than one dimension encounter severe difficulties if the points that approximations are based on are not regularly placed (grid based). Difficulties in solving differential equations often come from boundaries that may be irregularly shaped and from mixtures of scales that may require spatially variable resolution. It transpires that RBFs can replace (or supplement) polynomials in such situations, again leading to highly effective and accurate FD-type approximations.
6. **FD in the Complex Plane:** Measurable physical quantities do not involve complex numbers.[1] However, with most standard and special functions in the applied sciences being *analytic functions*, both mathematical analysis and computational procedures can benefit greatly from exploiting this feature. While such mathematical tools have seen much use during the last couple of centuries, the realization is far more recent that FD methods in the complex plane can also be remarkably effective.
7. **FD-based Methods for Quadrature and Infinite Sums:** This is again an area where commonly used methods on equispaced grids (the setting in which data is often available if not created just for the purpose of quadrature) relate closely to FD approximations. Complex plane FD approximations can be used for highly accurate contour integration of analytic functions.
8. **Fractional-Order Derivatives:** Although their history is nearly as long as that of regular (integer-order) derivatives, their range of applications has increased dramatically in recent decades. High-order accurate FD methods for their approximation amount to a recent development.

The main chapters are followed by seven appendices with supporting back-

[1] unless possibly in the context of quantum mechanics.

ground materials, also focused on heuristic insights and application practicalities.

Audience for Whom the Book Is Written

This book is primarily aimed toward students and researchers who are interested or actively engaged in scientific computing. For most of the material, no more background is needed than basic calculus and some knowledge of numerical analysis, linear algebra, and complex variables. This book is not intended as a sole textbook for an introductory numerical analysis or numerical differential equations course, but more for a graduate course aimed at providing the supplementary insights and perspectives that are essential for a good understanding, but which too often fall in the cracks between traditional course materials. This book will also provide educators with additional perspectives to those focused on in many textbooks.

Some General Remarks

Books with extensive mathematical content need to find a balanced path between rigor and heuristics. We tilt here quite strongly in favor of the latter, as this much more closely reflects how actual scientific computing is designed and carried out. The goal has been to present the relevant materials, not in some form of cookbook fashion, but instead to highlight the essential concepts behind different cost-effective FD-type computational opportunities. Another aspect for which a balanced path needs to be found is how much background material to include – that is, between presenting in a too terse fashion versus bloating the manuscript with topics that many readers may find unnecessary. In this regard, we tilt somewhat toward the former. In some cases, we only alert readers to computational opportunities, leaving details to cited references and to the appendices.

When developing a numerical solution strategy and a code for an application, it is often practical to start with low-order approximations, to most easily "get into the business." It will then often transpire that higher computational efficiency is needed, calling for the second step of upgrading to higher-order accurate approximations. A goal for this book is to provide perspectives for this second step.

Acknowledgments: My personal interest in the topics described in this book evolved from my PhD work at Uppsala University, supervised by Professor Heinz-Otto Kreiss. Part of my thesis was concerned with how the materials here in Chapter 1 lead toward those in Chapter 2 (i.e., how increasing-order accurate FD methods give rise to PS methods). For past professional influences and inspirations, I want also to acknowledge Professors Herbert B. Keller, Robert D. Richtmyer, Philip G. Saffman, and Gerald B. Whitham. I also thank Professor Nick Trefethen for enthusiastically reading and commenting on an early draft of this book. Most importantly, I thank my wife, Dr. Natasha Flyer, for all her love, support, and encouragement.

Software used: The manuscript was prepared in LyX (a front-end system to LaTeX).[2] The graphics and the computational results have mostly been obtained using MATLAB,[3] in some cases supplemented by the Advanpix[4] extended precision toolbox. In a few cases, brief Mathematica[5] codes have been included.

[2] LyX – The Document Processor, www.lyx.org/.
[3] MathWorks Inc., Natick, Massachusetts: www.mathworks.com.
[4] Multiprecision Computing Toolbox, Advanpix LLC, Yokohama, Japan, www.advanpix.com/.
[5] Wolfram Research, Inc., Champaign, IL.

1

Introduction to Finite Difference Methods

1.1 A Few Historical Notes

Finite difference (FD)-type discrete approximations can be traced back much earlier than when Gottfried Leibniz[1] and Isaac Newton[2] gave the first descriptions of calculus (in 1684 and 1687, respectively). The introduction of FDs (at first for interpolation) is often attributed to Jost Bürgi,[3] around 1592. Finite difference formulas of high orders of accuracy (especially for integration) were used by James Gregory in 1670.[4] Significant further early FD perspectives were provided by Isaac Newton and later by Brook Taylor[5] in 1715. Finite difference formulas were quite widely used for solving ordinary differential equations (ODEs) in the nineteenth century (notably for problems in fluid mechanics and for planetary orbit calculations). The pioneering work on the use of FD for partial differential equations (PDEs) dates back to the study by Lewis F. Richardson (1911)[6] (on potentially dangerous stresses in the first dam over the Nile in Aswan).

1.2 Finite Difference Formulas

The standard definition of the first derivative

$$f'(x) = \lim_{h \to 0} \frac{f(x+h) - f(x)}{h} \tag{1.1}$$

[1] 1646–1716, German mathematician, philosopher, scientist, and diplomat.

[2] 1643–1727, English mathematician, physicist, astronomer, alchemist, and theologian; widely recognized as one of the greatest scientists of all time.

[3] 1552–1632, Swiss mathematician, also a maker of clocks and astronomical instruments.

[4] Described further in Section 7.3.

[5] 1685–1731, English mathematician, of Taylor series fame.

[6] 1881–1953, English mathematician and pioneer of numerical weather forecasting.

is a simple example of an FD formula. Taylor expansion of (1.1) shows that

$$\frac{f(x+h)-f(x)}{h}=f'(x)+\frac{h}{2!}f''(x)+\frac{h^2}{3!}f'''(x)+\cdots=f'(x)+O(h^1), \quad (1.2)$$

that is, the approximation $f'(x)\approx\frac{f(x+h)-f(x)}{h}$ is accurate to *first order* (at the location x). The FD *weights* at the *nodes* x and $x+h$ can be abbreviated as $[-1\ \ 1]\ /\ h$. We will soon find it convenient to sketch FD *stencils* graphically. In this case:

$$\begin{array}{ccl} \circ & & \leftarrow \text{ entry for } f',\text{ weight } \{1\}, \\ \square & \square & \leftarrow \text{ entries for } f,\text{ weights } \{-\frac{1}{h},\ \frac{1}{h}\}, \\ \uparrow & \uparrow & \\ x & x+h & \leftarrow \text{ spatial locations.} \end{array} \quad (1.3)$$

The circle indicates the location in x for a derivative entry, and the squares indicate x-locations for function values (vertical spacing is here of no significance). Each of these locations has a *weight* associated with it. While the simplicity of this approximation is convenient (it uses only two adjacent function values to produce a derivative approximation), its low order of accuracy (first order; exact only for polynomials up through degree one) makes it almost entirely useless for practical computing.

A significantly better approximation to the first derivative is provided by

$$f'(x)=\frac{-\frac{1}{2}f(x-h)+\frac{1}{2}f(x+h)}{h}+O(h^2). \quad (1.4)$$

Figure 1.1 illustrates the two FD approximations (1.2) and (1.4). Although both give the same approximation for the tangent slope in the limit $h\to 0$, it is clear also visually that the second-order approximation is the more accurate one for small but nonzero h.

Taylor expansions provide a helpful means of verifying the order of accuracy if an FD formula is proposed. However, there are numerous more convenient ways to generate such formulas, as described in Section 1.2.1. Since the value of x in (1.2) and (1.4) – and in general for FD formulas – has no influence on the weights, we simplify the notation in Section 1.2.1 by approximating derivatives at $x=0$.

1.2.1 Some Algorithms for Generating FD Weights

In all but the last of the five methods described in Sections 1.2.1.1–1.2.1.5, the independent variable x can just as well be a complex variable z (with node points distributed in the complex plane rather than along the real axis).

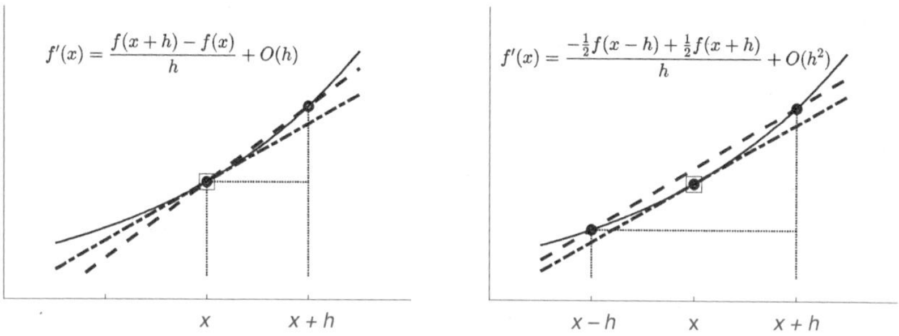

Figure 1.1 Graphical comparison between the first- and second-order approximations for $f'(x)$, as given by (1.2) and (1.4), respectively. The slope of the tangent lines (dash-dotted) corresponds to the true derivative value at the location x, and the slopes of the secant lines (dashed) correspond to the respective approximations.

1.2.1.1 FD Weights by Use of Monomial Test Functions

This algorithm is very flexible. Let L be any derivative operator (such as $\frac{d}{dx}$ and $\frac{d^2}{dx^2}$). We require a stencil with n nodes, located at $x_1, x_2, \ldots, x_n$ (distinct) to be exact for the first n monomials $1, x, x^2, \ldots, x^{n-1}$. This requirement can be written as

$$\begin{bmatrix} 1 & 1 & \cdots & 1 \\ x_1 & x_2 & \cdots & x_n \\ \vdots & \vdots & & \vdots \\ x_1^{n-1} & x_2^{n-1} & \cdots & x_n^{n-1} \end{bmatrix} \begin{bmatrix} w_1 \\ w_2 \\ \vdots \\ w_n \end{bmatrix} = \begin{bmatrix} L\,1|_{x=0} \\ L\,x|_{x=0} \\ \vdots \\ L\,x^{n-1}|_{x=0} \end{bmatrix}. \tag{1.5}$$

The weights w_k at nodes x_k, $k = 1, 2, \ldots, n$, can now be obtained by solving this linear system. The matrix is of Vandermonde type – often ill-conditioned, but never singular assuming the x_k are distinct.[7]

Verification of nonsingularity: The result is true for $n = 1$, in which case the matrix is $[1]$. Applying induction over n, assume the result is true up through size $n - 1$, and consider the size n case. Call temporarily $x_n = x$ and the matrix $A(x)$. Expanding along the last column, $\det(A(x))$ becomes a polynomial in x of degree $n - 1$, not identically zero (as, by the induction hypothesis, the coefficient for x^{n-1} is nonzero). All its $n - 1$ roots are accounted for by $x = x_1$, $x = x_2, \ldots, x = x_{n-1}$. It must therefore be nonzero for $x = x_n$.

Example 1.2.1 Create an FD formula that approximates $f''(0)$ based on function values $f(x_1)$ and $f(x_2)$ and first derivative values $f'(x_3)$ and $f'(x_4)$.

[7] A different proof to the one immediately following is given in Section A.2.

Solution The monomial test function algorithm generalizes immediately to cases with mixed types of input data. Enforcing

$$f''(0) \approx w_1 f(x_1) + w_2 f(x_2) + w_3 f'(x_3) + w_4 f'(x_4)$$

to be exact for the functions $1, x, x^2, x^3$ gives the following linear system to solve for the weights:

$$\begin{bmatrix} 1 & 1 & 0 & 0 \\ x_1 & x_2 & 1 & 1 \\ x_1^2 & x_2^2 & 2x_3 & 2x_4 \\ x_1^3 & x_2^3 & 3x_3^2 & 3x_4^2 \end{bmatrix} \begin{bmatrix} w_1 \\ w_2 \\ w_3 \\ w_4 \end{bmatrix} = \begin{bmatrix} 0 \\ 0 \\ 2 \\ 0 \end{bmatrix}.$$

The determinant of the coefficient matrix becomes

$$(x_1 - x_2)(x_3 - x_4)(2x_1^2 + 2x_1x_2 + 2x_2^2 - 3x_1x_3 - 3x_2x_3 - 3x_1x_4 - 3x_2x_4 + 6x_3x_4).$$

The first two factors tell that the system is singular in the case of coinciding nodes $x_1 = x_2$ or $x_3 = x_4$. The remaining quadratic form is not positive or negative definite,[8] implying possibilities of additional singularities for certain node locations. □

1.2.1.2 FD Weights by the Method of Exponential Test Functions

This method leads to the same linear system (1.5). However, it simplifies certain tasks (cf. Sections 7.5.1 and 8.3.1). We now apply L to the test function $e^{\xi x}$ rather than to monomials in x. Equating leading Taylor expansion terms in ξ in $\sum_{k=1}^{n} w_k e^{\xi x_k} = L\, e^{\xi x}\big|_{x=0}$ becomes, when written out more explicitly,

$$\left\{ \begin{array}{l} w_1 \left(1 + \xi x_1 + \frac{\xi^2 x_1^2}{2!} + \cdots \right) + \\ w_2 \left(1 + \xi x_2 + \frac{\xi^2 x_2^2}{2!} + \cdots \right) + \\ \qquad \vdots \\ w_n \left(1 + \xi x_n + \frac{\xi^2 x_n^2}{2!} + \cdots \right) \end{array} \right\} = L \left(1 + \xi x + \frac{\xi^2 x^2}{2!} + \cdots \right)\Bigg|_{x=0},$$

again giving the linear system (1.5).

[8] Seen, for example, by noting that it is linear in x_4 (and x_3). For any choice of x_1, x_2, x_3, it evaluates to zero for $x_4 = \frac{2x_1^2+2x_1x_2+2x_2^2-3x_1x_3-3x_2x_3}{3(x_1+x_2-2x_3)}$. More generally, writing it as

$$[x_1, x_2, x_3, x_4] \begin{bmatrix} 2 & 1 & -3/2 & -3/2 \\ 1 & 2 & -3/2 & -3/2 \\ -3/2 & -3/2 & 0 & 3 \\ -3/2 & -3/2 & 3 & 0 \end{bmatrix} \begin{bmatrix} x_1 \\ x_2 \\ x_3 \\ x_4 \end{bmatrix},$$

it is seen from the fact that the (symmetric) matrix has both positive and negative eigenvalues ($\lambda_1 = 6, \lambda_2 = 1, \lambda_3 = 0, \lambda_4 = -3$). However, if $x_3 = x_1$ and $x_4 = x_2$ (then a case of Hermite interpolation, cf. Section 1.4.3), the quadratic form evaluates to $-(x_1 - x_2)^2$, ensuring nonsingularity for $x_1 \neq x_2$.

1.2.1.3 FD Weights by Lagrange's Interpolation Formula

This approach is good for theoretical insight but less convenient to code. Consider, for example, deriving (1.4). The *Lagrange interpolation polynomial*[9] $p(x)$ that takes the values $f(-h), f(0), f(h)$ at $x = -h, 0, h$ is

$$p(x) = \frac{(x-0)(x-h)}{(-h-0)(-h-h)} f(-h) + \frac{(x+h)(x-h)}{(0+h)(0-h)} f(0) + \frac{(x+h)(x+0)}{(h+h)(h+0)} f(h).$$

A quick calculation shows that $p'(0) = \frac{-\frac{1}{2}f(x-h)+\frac{1}{2}f(x+h)}{h}$, in agreement with (1.4).

1.2.1.4 FD Weights by Recursion

This algorithm is computationally very fast and has excellent numerical stability. It follows from Lagrange's interpolation formula – see Fornberg (1988a) for details. We give it here only in the form of the MATLAB code shown in Algorithm 1.1. For example, the statement **weights(0, −2 : 2, 6)** returns the output

```
     0          0       1.0000        0          0
  0.0833    -0.6667       0        0.6667    -0.0833
 -0.0833     1.3333   -2.5000      1.3333    -0.0833
 -0.5000     1.0000       0       -1.0000     0.5000
  1.0000    -4.0000    6.0000     -4.0000     1.0000
     0          0          0          0          0
     0          0          0          0          0
```

The top line gives the weights for interpolation, which here is trivial as the interpolation point coincides with one of the grid points. The next four rows give the optimal weights for five-node approximations to derivatives of orders 1 to 4.[10] The last two rows are all zero, as derivatives of orders 5 and higher cannot be approximated based on just five nodes.

1.2.1.5 FD Weights by Padé-Based Algorithm

This algorithm was introduced in Fornberg (1998). Its derivation is a straightforward generalization of the special case described in Example 1.2.5. It requires equispaced nodes but is otherwise very general. It readily applies also to *implicit stencils*,[11] and it naturally uses exact rather than floating-point arithmetic. In the display form introduced in (1.3), we consider now stencils of the more general form

[9] Described in Appendix A, with illustration in Figure 1.3.
[10] Thus, lines 2 and 3 match the second row in Tables 1.1 and 1.2, respectively.
[11] Also described as *compact stencils*; see also Sections 3.3.2 and 4.3.1.1.

Algorithm 1.1 A MATLAB implementation of the weights algorithm in Section 1.2.1.4.

```
function c = weights(z,x,m)
% Calculates FD weights.
% Input parameters:
%  z Location where approximations are to be accurate,
%  x Row vector with x-coordinates for grid points (distinct but
%    otherwise arbitrarily located),
%  m Highest derivative that we want to find weights for
% Output parameter:
%  c Matrix, size (m+1,length(x)), containing in successive rows the
%    weights for derivatives 0,1,...,m.

n = length(x); c = zeros(m+2,n); c(2,1) = 1; x1 = repmat(x,n,1);
A = x1.'-x1;
b = cumprod([ones(n,1),A],2); rm = cumsum(ones(m+2,n-1))-1;
d = diag(b); d(1:n-1) = d(1:n-1)./d(2:n);
for i = 2:n
   mn = min(i,m+1);
   c(2:mn+1,i) = d(i-1)*(rm(1:mn,1).*c(1:mn,i-1)-(x(i-1)-z)* ...
   c(2:mn+1,i-1));
   c(2:mn+1,1:i-1) = ((x(i)-z)*c(2:mn+1,1:i-1)-rm(1:mn,1:i-1).* ...
   c(1:mn,1:i-1))./(x(i)- x1(1:mn,1:i-1));
end
c(1,:) = [];
end
```

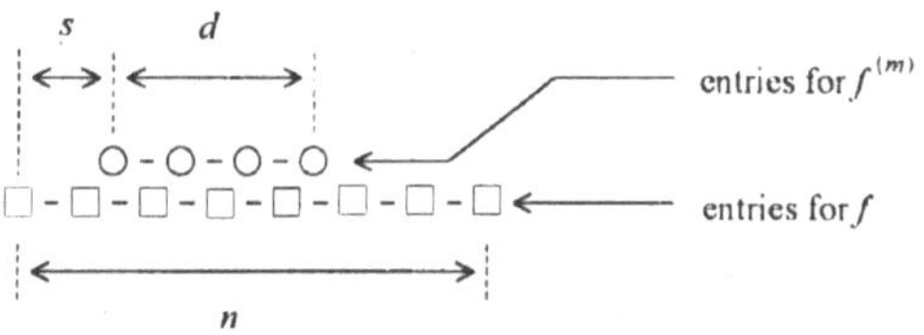

relating weights for the mth derivative at $d + 1$ locations spaced h apart with weights for the function at $n + 1$, also h-spaced locations, with the two sets shifted sideways by s units of h, where s need not be an integer. In symbolic languages, such as Mathematica here, just two lines of code suffice[12]:

```
t = PadeApproximant[x^s(Log[x]/h)^m,{x,1,{n,d}}];
CoefficientList[{Denominator[t],Numerator[t]},x]
```

The following examples illustrate how this algorithm can be used and (Example 1.2.5) also the approach for deriving it.

[12] The present usage of Padé expansions differs significantly from their more common applications to convergence acceleration and numerical analytic continuation of both Taylor and asymptotic expansions; see Section E.3 and, for example, Bender and Orszag (1978), Sections 8.3–8.6; Fornberg and Piret (2020), Section 3.2.9; and Trefethen (2013), Chapter 27.

Example 1.2.2 Find the weights in a stencil of the shape $\begin{smallmatrix} & \circ & \\ \square & \square & \square \end{smallmatrix}$ for approximating the second derivative.

Solution For $s = 1,\ d = 0,\ n = 2,\ m = 2$, the algorithm returns

$$\left\{\left\{h^2\right\}, \{1, -2, 1\}\right\},$$

corresponding to the explicit second-order accurate formula for the second derivative

$$f''(x) \approx \{f(x-h) - 2f(x) + f(x+h)\} \frac{1}{h^2}. \tag{1.6}$$

□

Example 1.2.3 Find the weights in a stencil of the shape $\begin{smallmatrix} \circ & \circ & \circ \\ \square & \square & \square \end{smallmatrix}$, again for approximating the second derivative.

Solution For $s = 0,\ d = 2,\ n = 2,\ m = 2$, the algorithm returns

$$\left\{\left\{\frac{h^2}{12}, \frac{5h^2}{6}, \frac{h^2}{12}\right\}, \{1, -2, 1\}\right\},$$

corresponding to the compact (implicit) fourth-order accurate formula for the second derivative

$$\begin{aligned} &\frac{1}{12}f''(x-h) + \frac{5}{6}f''(x) + \frac{1}{12}f''(x+h) \\ &\qquad \approx \{f(x-h) - 2f(x) + f(x+h)\} \frac{1}{h^2}. \end{aligned} \tag{1.7}$$

□

Example 1.2.4 Find the weights in a stencil of the shape $\begin{smallmatrix} \circ & \circ & \circ & \\ & & \square & \square \end{smallmatrix}$ for approximating the first derivative.

Solution For $s = -2,\ d = 2,\ n = 1,\ m = 1$, the output

$$\left\{\left\{\frac{5h}{12}, -\frac{4h}{3}, \frac{23h}{12}\right\}, \{-1, 1\}\right\}$$

is readily rearranged into

$$f(x+h) = f(x) + \frac{h}{12}(23f'(x) - 16f'(x-h) + 5f'(x-2h)). \tag{1.8}$$

We later (in Section 3.2.3) encounter this formula as the third-order Adams–Bashforth method for solving ODEs. □

Example 1.2.5 For Example 1.2.3, obtain the weights by explicitly carrying out the exponential test function approach.

Algorithm 1.2 A MATLAB implementation of the weights algorithm in Section 1.2.1.5.

```
function [w_f,w_der] = weights_Pade(s,d,n,m)
% Input parameters (type double or symbolic); d,n,m must be integers;
% If s is non-integer, symbolic form for s ensures exact arithmetic.
%  s,d,n  Define stencil shape
%      m  Order of derivative approximated
% Output parameters (symbolic variables)
%    w_f  Weight(s) for function values
%  w_der  Weight(s) for derivative values
syms x h;  [N,D] = numden(pade(x^s*log(x)^m,x,1,'Order',[n,d]));
w_f = fliplr(coeffs(N,'All'));  w_der = fliplr(coeffs(D,'All'))*h^m;
end
```

Solution The approximation should be of the form

$$b_{-1}f''(x-h)+b_0f''(x)+b_1f''(x+h)\approx c_{-1}f(x-h)+c_0f(x)+c_1f(x+h).$$

With $f=e^{\xi x}$, this becomes

$$\xi^2\left(b_{-1}e^{-\xi h}+b_0+b_1e^{\xi h}\right)e^{\xi x}\approx\left(c_{-1}e^{-\xi h}+c_0+c_1e^{\xi h}\right)e^{\xi x}.$$

After canceling $e^{\xi x}$ and substituting $e^{\xi h}=s$ (i.e., $\xi=\frac{1}{h}\log s$), this can be written as

$$\left\{\frac{1}{h}\log s\right\}^2\approx\frac{c_{-1}+c_0s+c_1s^2}{b_{-1}+b_0s+b_1s^2}. \tag{1.9}$$

This relation should be as accurate as possible for $\xi\to 0$, that is, for $s\to 1$. The Padé expansion (cf. Section E.3) of $\left\{\frac{1}{h}\log s\right\}^2$ around $s=1$ with numerator and denominator degrees both equal to 2 becomes

$$\left\{\frac{1}{h}\log s\right\}^2\approx\frac{1}{h^2}\,\frac{(s-1)^2}{1+(s-1)+\frac{1}{12}(s-1)^2}=\frac{12-24s+12s^2}{h^2(1+10s+s^2)}.$$

Equating coefficients with (1.9) gives (1.7). □

The MATLAB code in Algorithm 1.2 implements this Padé-based algorithm using the Symbolic Toolbox. For example, to compute the weights in the second row in Table 4.1, the statement

`[w_f, w_der] = weights_Pade(sym(3/2), 0, 3, 1)`

produces the output `w_f = [1, -27, 27, -1]`, `w_der = 24 * h`.

Table 1.1 *Weights for centered FD approximations of the first derivative on an equispaced grid (omitting the factor 1/h).*

Order						Weights	First derivative						
2						$-\frac{1}{2}$	0	$\frac{1}{2}$					
4					$\frac{1}{12}$	$-\frac{2}{3}$	0	$\frac{2}{3}$	$-\frac{1}{12}$				
6				$-\frac{1}{60}$	$\frac{3}{20}$	$-\frac{3}{4}$	0	$\frac{3}{4}$	$-\frac{3}{20}$	$\frac{1}{60}$			
8			$\frac{1}{280}$	$-\frac{4}{105}$	$\frac{1}{5}$	$-\frac{4}{5}$	0	$\frac{4}{5}$	$-\frac{1}{5}$	$\frac{4}{105}$	$-\frac{1}{280}$		
10		$-\frac{1}{1260}$	$\frac{5}{504}$	$-\frac{5}{84}$	$\frac{5}{21}$	$-\frac{5}{6}$	0	$\frac{5}{6}$	$-\frac{5}{21}$	$\frac{5}{84}$	$-\frac{5}{504}$	$\frac{1}{1260}$	
⋮		↓	↓	↓	↓	↓	⋮	↓	↓	↓	↓	↓	
Limit	⋯	$-\frac{1}{5}$	$\frac{1}{4}$	$-\frac{1}{3}$	$\frac{1}{2}$	-1	0	1	$-\frac{1}{2}$	$\frac{1}{3}$	$-\frac{1}{4}$	$\frac{1}{5}$	⋯

1.2.2 Some Tables of FD Formulas

Especially with the recursion and Padé algorithms (in Sections 1.2.1.4 and 1.2.1.5, respectively), it is very easy to generate tables of FD weights. Four examples are given in Tables 1.1–1.4.[13]

We can make a number of observations from these relating to increasing orders of accuracy:

i. For centered FD approximations, the weights converge to well-defined limits – derived later in Section 2.1 for an arbitrary-order derivative.

ii. As seen for the first and second derivatives in Tables 1.1 and 1.2, and for a general-order derivative in (2.1), the weights for centered FD approximations decay in magnitude with distance k from the stencil center at the rate of $O(1/k)$ for odd-order derivatives, and $O(1/k^2)$ for those of even order. These slow decay rates are in some contexts problematic, since analytically, a derivative is a local property of a function and should not depend heavily on distant data.[14]

iii. For one-sided approximations, the weights diverge rapidly with increasing orders of accuracy (as seen in Tables 1.3 and 1.4). Ways to understand and control this will be discussed in several contexts later in this book (including Sections 1.4, 5.3.7, and F.3).

[13] A simple relation between the entries in Tables 1.1 and 1.2 is given in Section 6.3.3.

[14] This comment does not apply to FD in the complex plane (cf. Chapter 6).

Table 1.2 *Weights for centered FD approximations of the second derivative on an equispaced grid (omitting the factor $1/h^2$).*

Order		Weights	Second derivative									
2						1	-2	1				
4					$-\frac{1}{12}$	$\frac{4}{3}$	$-\frac{5}{2}$	$\frac{4}{3}$	$-\frac{1}{12}$			
6				$\frac{1}{90}$	$-\frac{3}{20}$	$\frac{3}{2}$	$-\frac{49}{18}$	$\frac{3}{2}$	$-\frac{3}{20}$	$\frac{1}{90}$		
8			$-\frac{1}{560}$	$\frac{8}{315}$	$-\frac{1}{5}$	$\frac{8}{5}$	$-\frac{205}{72}$	$\frac{8}{5}$	$-\frac{1}{5}$	$\frac{8}{315}$	$-\frac{1}{560}$	
10		$\frac{1}{3150}$	$-\frac{5}{1008}$	$\frac{5}{126}$	$-\frac{5}{21}$	$\frac{5}{3}$	$-\frac{5269}{1800}$	$\frac{5}{3}$	$-\frac{5}{21}$	$\frac{5}{126}$	$-\frac{5}{1008}$	$\frac{1}{3150}$
$\vdots$		$\downarrow$	$\downarrow$	$\downarrow$	$\downarrow$	$\downarrow$	$\vdots$	$\downarrow$	$\downarrow$	$\downarrow$	$\downarrow$	$\downarrow$
Limit	$\cdots$	$\frac{2}{5^2}$	$-\frac{2}{4^2}$	$\frac{2}{3^2}$	$-\frac{2}{2^2}$	$\frac{2}{1^2}$	$-\frac{\pi^2}{3}$	$\frac{2}{1^2}$	$-\frac{2}{2^2}$	$\frac{2}{3^2}$	$-\frac{2}{4^2}$	$\frac{2}{5^2}$ $\cdots$

Table 1.3 *Weights for one-sided FD approximations of the first derivative on an equispaced grid (omitting the factor $1/h$).*

Order			Weights	First derivative							
1	-1	1									
2	$-\frac{3}{2}$	2	$-\frac{1}{2}$								
3	$-\frac{11}{6}$	3	$-\frac{3}{2}$	$\frac{1}{3}$							
4	$-\frac{25}{12}$	4	-3	$\frac{4}{3}$	$-\frac{1}{4}$						
5	$-\frac{137}{60}$	5	-5	$\frac{10}{3}$	$-\frac{5}{4}$	$\frac{1}{5}$					
6	$-\frac{49}{20}$	6	$-\frac{15}{2}$	$\frac{20}{3}$	$-\frac{15}{4}$	$\frac{6}{5}$	$-\frac{1}{6}$				
7	$-\frac{363}{140}$	7	$-\frac{21}{2}$	$\frac{35}{3}$	$-\frac{35}{4}$	$\frac{21}{5}$	$-\frac{7}{6}$	$\frac{1}{7}$			
8	$-\frac{761}{280}$	8	-14	$\frac{56}{3}$	$-\frac{35}{2}$	$\frac{56}{5}$	$-\frac{14}{3}$	$\frac{8}{7}$	$-\frac{1}{8}$		
9	$-\frac{7129}{2520}$	9	-18	28	$-\frac{63}{2}$	$\frac{126}{5}$	-14	$\frac{36}{7}$	$-\frac{9}{8}$	$\frac{1}{9}$	
10	$-\frac{7381}{2520}$	10	$-\frac{45}{2}$	40	$-\frac{105}{2}$	$\frac{252}{5}$	-35	$\frac{120}{7}$	$-\frac{45}{8}$	$\frac{10}{9}$	$-\frac{1}{10}$

1.3 Errors When Applying FD Formulas

Two types of errors arise when applying FD approximations to a function:

1. **Truncation errors:** The FD formula has an error of size $O(h^p)$, where p is the approximation's order of accuracy.
2. **Rounding errors:** Typical double precision accuracy in a computer has a relative error level of around 10^{-16}.

Table 1.4 *Weights for one-sided FD approximations of the second derivative on an equispaced grid (omitting the factor* $1/h^2$*).*

Order	Weights	Second derivative								
1	1	-2	1							
2	2	-5	4	-1						
3	$\frac{35}{12}$	$-\frac{26}{3}$	$\frac{19}{2}$	$-\frac{14}{3}$	$\frac{11}{12}$					
4	$\frac{15}{4}$	$-\frac{77}{6}$	$\frac{107}{6}$	-13	$\frac{61}{12}$	$-\frac{5}{6}$				
5	$\frac{203}{45}$	$-\frac{87}{5}$	$\frac{117}{4}$	$-\frac{254}{9}$	$\frac{33}{2}$	$-\frac{27}{5}$	$\frac{137}{180}$			
6	$\frac{469}{90}$	$-\frac{223}{10}$	$\frac{879}{20}$	$-\frac{949}{18}$	41	$-\frac{201}{10}$	$\frac{1009}{180}$	$-\frac{7}{10}$		
7	$\frac{29531}{5040}$	$-\frac{962}{35}$	$\frac{621}{10}$	$-\frac{4006}{45}$	$\frac{691}{8}$	$-\frac{282}{5}$	$\frac{2143}{90}$	$-\frac{206}{35}$	$\frac{363}{560}$	
8	$\frac{6515}{1008}$	$-\frac{4609}{140}$	$\frac{5869}{70}$	$-\frac{6289}{45}$	$\frac{6499}{40}$	$-\frac{265}{2}$	$\frac{6709}{90}$	$-\frac{967}{35}$	$\frac{3407}{560}$	$-\frac{761}{1260}$

1.3.1 Truncation Errors

These can be found theoretically by Taylor expanding the FD formula, as in (1.2). Numerically, one can evaluate the approximation and record the error. Then a log–log type plot showing $\log|\text{Error}|$ versus $\log h$ will feature (near-) straight lines with the slope p. These $O(h^p)$-size errors become smaller when h decreases.

1.3.2 Rounding Errors

In standard double precision arithmetic, both weights and function values are typically uncertain by $O(10^{-16})$. For FD formulas for the kth derivative, having a h^k in their denominator, rounding errors can be roughly estimated as $O(10^{-16})/h^k$, that is growing rapidly when h decreases.

1.3.3 Total Errors

The total error becomes smallest when the two error types match in size, that is, when $O(h^p) \approx O(10^{-16})/h^k$, occurring for $h \approx 10^{-16/(p+k)}$, in which case $|\text{Error}| \approx 10^{-16\left(\frac{p}{p+k}\right)}$. Figure 1.2 illustrates the schematic observations, by showing the error when $\left.\frac{d}{dx}e^{-x/2}\right|_{x=0} = \frac{1}{2}$ is approximated using the first four FD stencils in Table 1.1 (denoted according to their accuracy orders as FD2,. . . , FD8). Apart from illustrating the two error sources, the figure also gives a hint about why we in this book focus on higher-order FD methods. When solving differential equations, approximations are needed at all grid

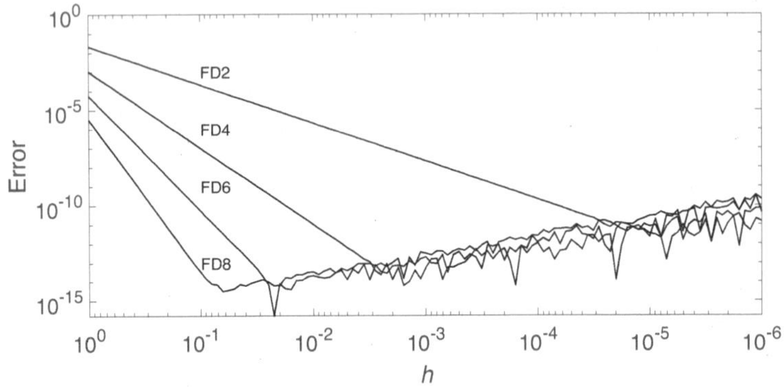

Figure 1.2 Errors when approximating $\frac{d}{dx}e^{-x/2}\big|_{x=0} = \frac{1}{2}$ with different grid spacings h and using the first four FD stencils in Table 1.1. The respective orders of accuracy are confirmed by the slopes of the initial straight-line trends. The jagged trend for smaller h reflect $O(10^{-16})/h$-sized rounding errors.

points across a domain. If we, for example, want to reach a 10^{-10} error level, the FD8 approximation (with four times as many weights as FD2) is seen to need only about 1/10,000 as many node points, typically making it vastly more cost-effective (already in 1-D; in 2-D, the factor would be 10^{-8}, etc.).

1.4 The Runge Phenomenon

High-degree polynomial interpolation with equispaced nodes is notorious for often being violently oscillatory near the end points of an interval. This plays a big role in the way that increasing order FD methods are designed.

1.4.1 Illustration of Lagrange Polynomials

Figure 1.3 illustrates the Lagrange polynomials $L_k(x)$ (see equation (A.1) in Appendix A) in the case of nodes located at $x_k = k$, $k = 0, 1, \ldots, 8$. None of these polynomials feature strikingly large spurious oscillations near the interval center, but several do that near the interval ends. Combining these kernel functions together according to the function values $f(x_k)$ will produce a polynomial interpolant sharing the same property, that is, being relatively smooth near the interval center, but with a high likelihood of oscillations near its ends.

Figure 1.3 also gives insight into the FD weight tables. The weights given in the next to last row of Tables 1.3 and 1.4 are exactly the first and second

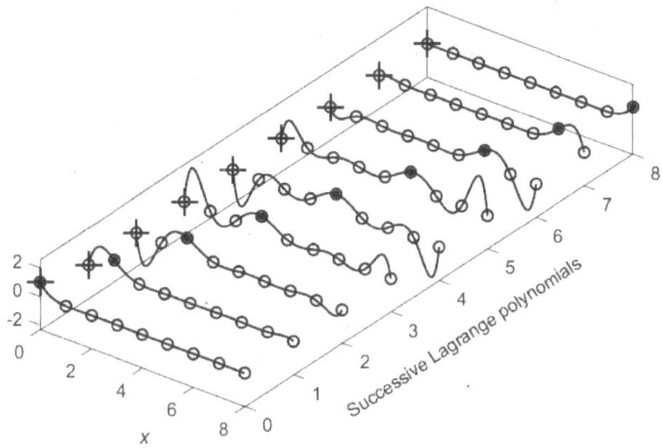

Figure 1.3 The Lagrange polynomials $L_k(x)$ in the case of $x_k = k$, $k = 0, 1, \ldots, 8$. The nodes at which they are zero are marked with open circles and the node at which the value is one is shown with a filled circle.

derivative of these shown curves at their left end point $x = 0$ (marked here by crosses). For the centered FD approximations (the next to last rows in Tables 1.1 and 1.2; again of stencil width of 9 nodes), the weights are similarly the matching derivatives of these curves at their center location.

1.4.2 Illustration of the Runge Phenomenon

Figure 1.4 shows a commonly illustrated case of the Runge phenomenon – interpolating the function

$$f(x) = \frac{1}{1 + 16x^2} \tag{1.10}$$

at equispaced nodes over $x \in [-1, 1]$. Without here going into the extensive literature on Runge phenomenon analysis,[15] let us just note some pertinent highlights that are relevant to the present goal of understanding its influence on FD (and PS) methods:

i. Figure 1.4 again illustrates why centered FD approximations are much more accurate than one-sided ones, as in both cases these return the exact derivative of interpolating polynomials.
ii. With N nodes, the *envelope* of the oscillations in the error (to leading order;

[15] The pioneering work by C. Runge (1901) has been recalled in, for example, Fornberg (1996) and Trefethen (2000, 2013). One important starting point is the formula (A.7).

$x \in [-1, 1]$) is given by $\alpha(x)^N$, where $\alpha(x) = e^{\phi(x)-c}$, and

$$\phi(x) = \frac{1}{2}\left((1-x)\log(1-x) + (1+x)\log(1+x)\right). \tag{1.11}$$

The function $f(x)$ that is interpolated enters only through the constant c, which in the present case of (1.10) becomes $c = \frac{1}{2}\log\frac{17}{16} + \frac{1}{4}\arctan 4 \approx 0.3618$.[16] Figure 1.5 shows the function $\alpha(x)$ for the present $f(x)$. We can read off that the interpolant converges exponentially fast in a central region (at $x = 0$ as 0.6964^N) and diverges exponentially toward the edges, with the transition points at $x \approx \pm 0.7942$ (independent of N).

iii. While interpolation generally is much more stable numerically than extrapolation (or approximations near to interval ends), the extreme severity of the Runge phenomenon is to a large extent due to the present use of polynomials as opposed to other types of interpolating functions.[17]

Although there is no simple universal remedy against the Runge phenomenon, various ways to reduce the damage it causes have been devised (such as the RBF-FD PHS+poly approach, described in Section 5.4.1 and, focusing on its properties at boundaries, in Section F.3).[18]

1.4.3 Hermite Interpolation

In certain applications, both $f(x_k)$- and $f'(x_k)$-values are available at all node points. Hermite interpolation can then be an alternative to Lagrange interpolation (again finding the unique lowest degree polynomial that fits the data). Finite difference weights to be simultaneously applied to both these sets can readily be obtained, for example, by generalizing the algorithms in Sections 1.2.1.1 and 1.2.1.4.[19] However, with N nodes, these polynomials will be of degree $2N - 1$ (rather than $N - 1$), off-setting possible advantages with respect to the Runge phenomenon. In the case illustrated in Figure 1.4, the break point between convergence and divergence is unchanged, again at $x \approx \pm 0.7942$; see Figure 1.6.

[16] Its value can be found by generalizing x to z complex and noting that convergence/divergence breakeven should occur at the most limiting singularity of $f(z)$; here at $z = \pm\frac{i}{4}$.

[17] For example, in contrast to rational functions, all polynomials $p(x)$ of high orders must diverge rapidly as $|x|$ increases.

[18] It is shown in Platte et al. (2011) that no stable algorithm can converge exponentially fast for all analytic functions if using only equispaced data on an interval within their region of analyticity. However, "root-exponential" convergence is possible (faster than any algebraic order).

[19] For details and MATLAB code, see Fornberg (2021a).

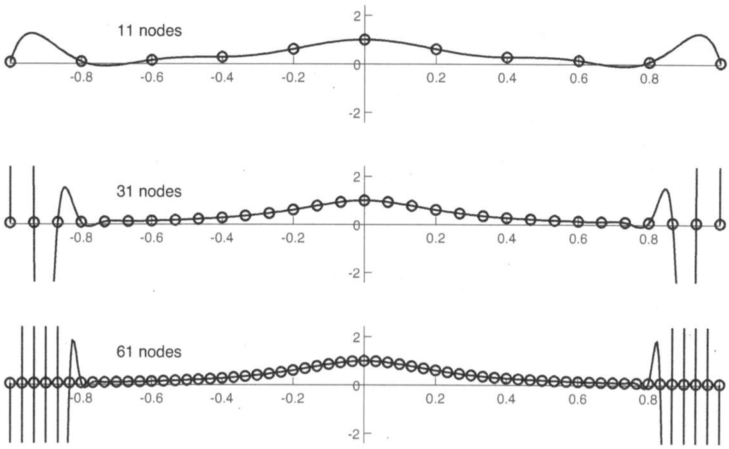

Figure 1.4 Equispaced polynomial interpolation of $f(x) = \frac{1}{1+16x^2}$ over $[-1, 1]$ using 11, 31, and 61 nodes.

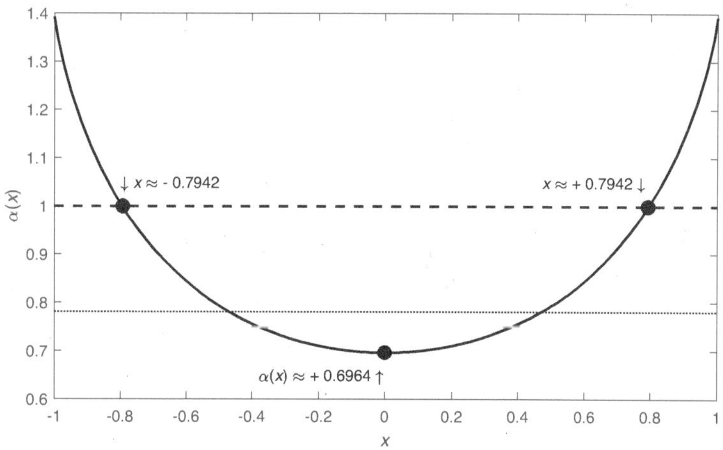

Figure 1.5 Solid curve: The function $\alpha(x)$ describing the Runge phenomenon oscillation envelope for equispaced interpolation over $[-1, 1]$ in the case of $f(x)$ given in (1.10). In particular, it always holds that $\alpha(\pm 1) - \alpha(0) = \log 2 \approx 0.6931$. Dotted straight line: The counterpart curve if the nodes are not equispaced but Chebyshev-distributed. For different functions $f(x)$, this curve and line get shifted up or down (with different amounts), but do not otherwise change shape.

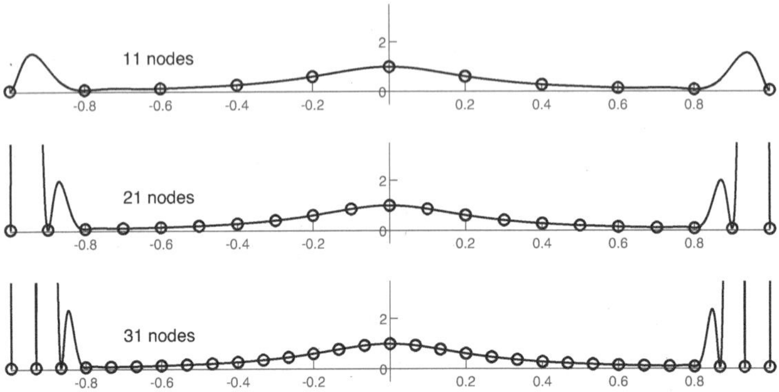

Figure 1.6 Equispaced Hermite interpolation of $f(x) = \frac{1}{1+16x^2}$ over $[-1, 1]$ using 11, 21, and 31 nodes.

1.4.4 Brief Comments on Chebyshev-Distributed Nodes

As seen in Tables 1.3 and 1.4, the Runge phenomenon is quite mild for small FD stencils, but then grows exponentially with the degree of the polynomials that the stencils are based on. Pseudospectral methods (Chapter 2) can be seen as pushing node numbers and polynomial degrees to extremely high values. As an alternative to an equispaced grid with nodes on $[-1, 1]$ located at

$$x_k = -1 + 2k/N, \qquad k = 0, 1, \ldots, N,$$

a commonly used Runge phenomenon remedy is then to cluster the nodes strongly toward the interval ends, as with the Chebyshev node distribution

$$x_k = -\cos(k\pi/N), \qquad k = 0, 1, \ldots, N.$$

If using approximations based on a single interpolating polynomial in the $N \to \infty$ limit, the dotted line in Figure 1.5 shows the counterpart $\alpha(x)$ curve, which in the case of (1.10) works out to become the constant $\alpha(x) \equiv \frac{1}{4}(\sqrt{17} - 1) \approx 0.7808$. The global polynomial interpolant now converges exponentially across $[-1, 1]$. Compared to equispaced nodes, improved accuracy near interval ends has been obtained in exchange for a loss of accuracy near the interval center.[20] With the focus in this book on FD approximations well short of the infinite order limit, the Runge phenomenon is often not a major issue.[21]

[20] The high accuracy seen near the interval center in Figure 1.5 explains the excellent accuracy of centered high-order FD approximations.

[21] Node clustering at boundaries can still be beneficial for other purposes, such as for achieving locally increased resolution (e.g., for resolving boundary layers) and for representing irregular boundary shapes. The generalization from grid-based FD to mesh-free RBF-FD discretizations is described in Chapter 5.

2

Brief Summary of Pseudospectral Methods

Already a casual glance at Tables 1.1 and 1.2, looking down their columns (especially the ones next to the center column) suggests that there exist simple infinite-order accurate limits. This limit process offers an intuitive approach to PS methods (in the absence of boundaries), supplementing the Fourier-based approach described in Section 2.1.

If boundaries are present, Tables 1.3 and 1.4 tell a completely different story, which was addressed more theoretically in Section 1.4. A commonly used way to address the divergence of weights with increasing orders of accuracy is to cluster the nodes heavily toward the domain boundaries – not because higher resolution may be needed there, but to control the polynomial Runge phenomenon. This approach has been extensively analyzed and explored computationally. It underlies classical topics, such as orthogonal polynomials and Gaussian quadrature, as well as more modern computational approaches for interpolation, functional representation, PS methods for ODEs and PDEs, and so on (Trefethen, 2013). Chebyshev polynomials are particularly important in this context and form the basis for the Chebfun software system.[1] Nevertheless, this falls outside the scope of the present book, as our focus is on high-order FD methods; local approximations, either on equispaced grids, or on irregular node layouts (however, only when such are required by an application rather than to control undesirable numerical artifacts). Sources for further reading on PS methods (with and without boundaries) include Boyd (2000); Fornberg (1996); Trefethen (2000).

Two reasons why we here consider equispaced PS methods (although they are not local approximations) are:

[1] Chebfun is a free/open-source software system written in MATLAB for numerical computation with functions of a real variable. It is based on the idea of overloading MATLAB's commands for vectors and matrices to analogous commands for functions and operators. Website: www.chebfun.org.

i. for insight about what happens when FD approximations are pushed toward extremely high orders of accuracy; and
ii. when discussing irregular geometries in Chapter 5, we will follow the opposite theme – radial basis functions (RBFs) historically originated in what we can view as their PS (global, spectrally accurate) case, and were later backtracked to provide local FD-type approximations (nowadays with the acronym RBF-FD).

As for the rest of this book, the brief summary on equispaced pseudospectral (PS) methods is more heuristic than rigorous, emphasizing aspects that relate to the two issues mentioned earlier.

2.1 Some PS Observations on Equispaced Nodes

2.1.1 PS Weights on an Infinite Interval

Each of the Tables 1.1 and 1.2 show as their bottom row the weights in the infinite order accurate PS limit. These weights are readily obtainable in closed form:

Theorem 2.1.1 *On a unit-spaced grid, the limit weight $w_k^{(m)}$ for the mth derivative at position k is given by*

$$w_k^{(m)} = \frac{d^m}{dx^m}\left(\frac{\sin \pi x}{\pi x}\right)\Bigg|_{x=k}. \tag{2.1}$$

Proof For a centered FD stencil over $[-N, N]$, the FD weight $w_k^{(m)}$ tells how much changing the function value $f(k)$ will change the mth derivative of the interpolating Lagrange polynomial at $x = 0$. With $L_N(x)$ denoting the Lagrange kernel

$$L_N(x) = \frac{(x+N)(x+N-1)\cdot \cdots \cdot (x+1)(x-1)\cdot \cdots \cdot (x-N+1)(x-N)}{(k+N)(k+N-1)\cdot \cdots \cdot (1)(-1)\cdot \cdots \cdot (k-N+1)(k-N)},$$

$w_k^{(m)} = \frac{d^m}{dx^m}(L_N(x))\big|_{x=k}$. It remains only to note that the sine-product formula $\frac{\sin \pi x}{\pi x} = \prod_{n=1}^{\infty}\left(1 - \frac{x^2}{n^2}\right)$ shows that $\lim_{N\to\infty} L_N(x) = \frac{\sin \pi x}{\pi x}$. □

2.1.2 PS Application on a Periodic Interval

It may seem that equispaced PS methods might be computationally impractical, as their stencil width is infinite. That need not be the case. If the function to be differentiated is periodic, one can imagine it extended to infinite width. The following then holds:

Theorem 2.1.2 *If a function $f(x)$ is periodic and given on an equispaced grid, the PS approximation to any derivative is exactly the same as if the function was interpolated with the lowest degree trigonometric polynomial, and this interpolant was then differentiated analytically.*

For a proof of this result, we refer to Fornberg (1987), Appendix A and Fornberg (1990), Appendix C.

Both the trigonometric interpolation and the return of the derivatives back to physical space is most easily done with a single Fast Fourier Transform (FFT – see Section C.4). Analytic differentiation in Fourier space simply amounts to the multiplication of each Fourier coefficient by its wave number (or a power of the same wave number in case of higher derivatives). While applying increasingly wide FD stencils incurs a cost proportional to the stencil width, the FFT drops the cost for the global PS stencils from $O(N^2)$ to $O(N \log N)$.

The name pseudospectral (as opposed to just spectral) is motivated by these transitions (for a time-dependent problem) forward and backward between physical and Fourier space at each time step (rather than remaining in Fourier space, and time-stepping Fourier coefficients). With the PS approach, one can effectively utilize both that differentiation is a simple algebraic operation in Fourier space and that variable coefficients and nonlinearities are similarly simple algebraic operations in physical space.

2.1.3 Accuracy of FD Approximations on a Periodic Grid

Smooth periodic functions (assuming 2π-periodicity for notational convenience) can be Fourier expanded, with the goal then to handle accurately as wide range as possible of Fourier modes $e^{i\,\omega x}$. With a grid spacing h, *aliasing* (cf. Section C.3.2) limits the range the grid can represent to $-\frac{\pi}{h} < \omega < +\frac{\pi}{h}$. Now

$$\frac{d}{dx} e^{i\,\omega x} = i\,\omega\, e^{i\,\omega x} \tag{2.2}$$

holds exactly. Approximating $\frac{d}{dx}$ with standard second-order FD gives instead

$$\text{FD2}(\,e^{i\,\omega x}) = \frac{e^{i\,\omega(x+h)} - e^{i\,\omega(x-h)}}{2h} = i\,\frac{\sin \omega h}{h}\, e^{i\,\omega x}, \tag{2.3}$$

generalizing (see Kreiss and Oliger, 1972) for order p (even) to

$$\text{FD}\{p\}(e^{i\,\omega x}) = i\,\frac{\sin \omega h}{h} \left\{ \sum_{k=0}^{p/2-1} \frac{(k!)^2}{(2k+1)!} \left(2 \sin \frac{\omega h}{2}\right)^{2k} \right\} e^{i\,\omega x}.$$

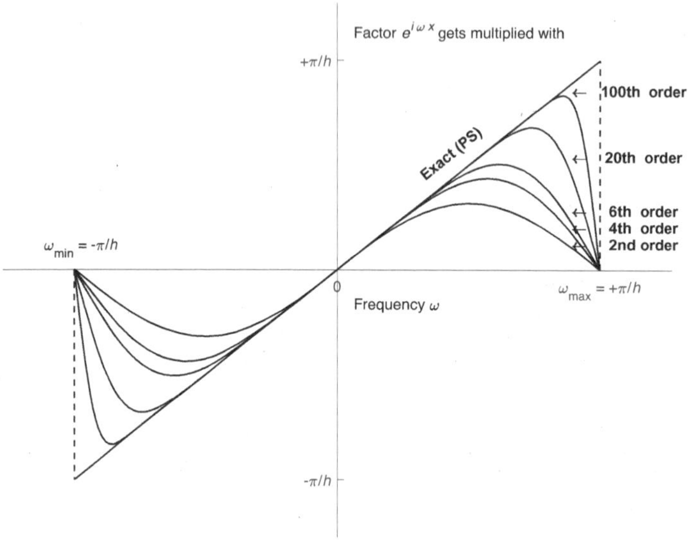

Figure 2.1 The factor a Fourier mode $e^{i\,\omega x}$ gets multiplied by (omitting a factor of i) when differentiated analytically versus when the differentiation is approximated by centered FD approximations of increasing orders, with grid spacing h.

Figure 2.1 shows the factor that multiplies $i\,e^{i\,\omega x}$ in the RHSs of (2.2) and (2.3) – as functions of ω, together with the corresponding factors in case of some additional accuracy orders.[2]

Example 2.1.1 Consider the PDE $\frac{\partial u}{\partial t} + \frac{\partial u}{\partial x} = 0$ over the periodic interval $x \in [0, 2\pi]$ from time $t = 0$ to $t = 40\pi$ (i.e., when the solution has traveled 20 times across the interval). Estimate the percentage of Fourier modes originally present on a grid with spacing h that have still retained some accuracy at the final time.

Solution At time t, the analytic solution for an initial mode $e^{i\,\omega x}$ is $e^{i\,\omega(x-t)}$. For example, using FD2, it follows from (2.3) that this mode has become $e^{i\,\omega x - i\frac{\sin \omega h}{h}t}$. Similarly, for each ω, we can read off the phase error in the mode as t multiplied by the vertical distance between the "Exact" and the corresponding FD curve in Figure 2.1. If that phase error has grown to π, the

[2] The curves for order p fit the exact (PS) curve much like truncated Taylor expansions, that is, with p derivatives correct at $\omega = 0$. This generally works well, as functions often have most of their "energy" in low modes. However, for applications where a certain frequency band is of primary interest (e.g., seismic exploration with band-limited forcing), FD schemes can be customized for that range (rather than optimized around) – see Section F.1. The information contained in Figure 2.1 has been used repeatedly to compare the effectiveness of different approximation orders in the context of solving wave-type PDEs (Fornberg, 1987; Kreiss and Oliger, 1972).

mode has the opposite sign to what it should have and is outright destructive to the accuracy. Below, we consider a mode to be accurate if this phase error is less than $\pi/4$. With this, we obtain the data shown in Table 2.1. □

The following are some conclusions we can draw from this example:

i. FD2 is extremely wasteful – of all Fourier modes initially present on the grid, over 90% get wasted already in the low accuracy case $N = 10$, and over 99% in the highest accuracy case of $N = 10,000$. These modes need to be included in FD2 computations but add nothing to the resulting accuracy. They have lost their connection with the actual PDE, becoming just spurious modes, which must be prevented from growing. While ensuring nongrowth is reasonably easy for constant coefficient linear PDEs, this can be a challenge for variable coefficient and nonlinear PDEs (Fornberg, 1973; Gustafsson et al., 2013).
ii. This issue with "wasted" modes gets much more severe in higher dimensions, as all numbers in Table 2.1 then need to be raised to the power of the number of dimensions, as illustrated for 3-D in Figure 2.2 (exemplifying the concept of "curse of dimensionality"). For example, in the moderate accuracy case of using $N = 100$ nodes in each of the three directions, FD2 has retained about $0.0425^3 \approx 0.000077$ of the modes initially present, versus for FD6 $0.2133^3 \approx 0.0097$. While this is still a small number, it is over 100 times larger than that for FD2.
iii. Increasing the FD accuracy order is highly beneficial for accuracy but, if pushed too far, comes with trade-offs (in terms of geometric flexibility, handling of boundary conditions, approximation quality for nonsmooth functions, etc.). High-order accurate FD occupies a sweet spot between computational inefficiency and these other issues.
iv. An extremely rough rule of thumb that's sometimes used is that a second-order method needs about 50 nodes within the smallest wavelength to be resolved and a fourth-order method about 7 (decreasing in the periodic PS limit to 2). Since many large-scale weather and geophysics models are of a relatively low order of accuracy, it is a gross misconception to equate their resolution ability with their distance between adjacent node points.

2.2 Gibbs Phenomenon

In the context of solving PDEs with variable coefficients, there may be discontinuities in either the coefficients or the solution (or both). The focus in

Table 2.1 *In Example 2.1.1, the proportion of the originally present Fourier modes that have retained accuracy at the final time* $t = 40\pi$ *when altogether using N nodes in space over the periodic 1-D interval* $x \in [0, 2\pi]$.

	N = 10	N = 100	N = 1,000	N = 10,000
PS	1.0000	1.0000	1.0000	1.0000
FD100	0.7987	0.7595	0.7270	0.6981
FD50	0.7184	0.6649	0.6209	0.5824
FD20	0.5692	0.4929	0.4326	0.3819
FD10	0.4226	0.3325	0.2261	0.2141
FD6	0.3043	0.2133	0.1524	0.1093
FD4	0.2143	0.1317	0.0828	0.0522
FD2	0.0947	0.0425	0.0197	0.0091

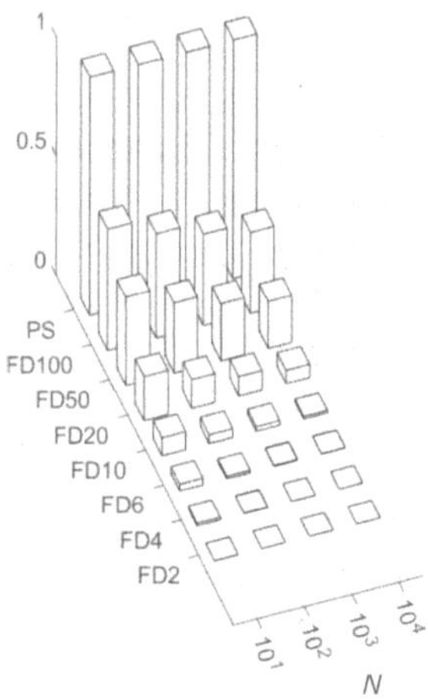

(a)

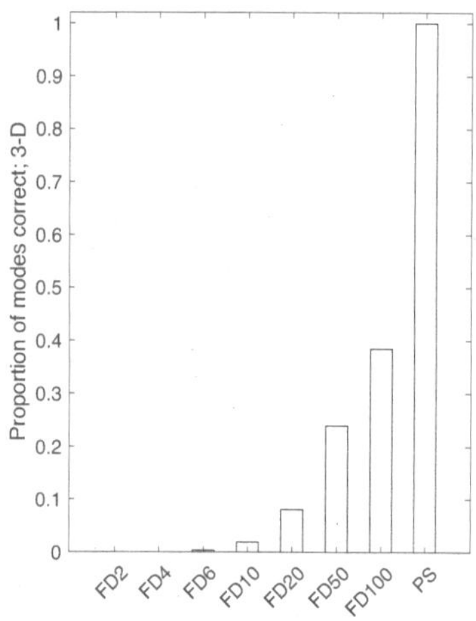

(b)

Figure 2.2 Graphical counterpart to Table 2.1, but for 3-D, that is, showing the cube of the numbers in the table. (a) Different order methods for different numbers of nodes along each direction, (b) The case of $N = 1,000$ nodes in each direction.

this present section is limited to introducing an issue that arises whenever one attempts to approximate a discontinuous function $u(x)$ using only smooth basis functions. This situation was first studied in the context of Fourier expansions, and it gives rise to the *Gibbs phenomenon*.[3] Figure 2.3 illustrates three cases:

[3] Also known as the Gibbs–Wilbraham phenomenon; discovered and analyzed by H. Wilbraham (1848) and rediscovered about half a century later, first in some experiments with a mechanical Fourier analyzer constructed by A. Michelson and S. W. Stratten, and after that, studied again

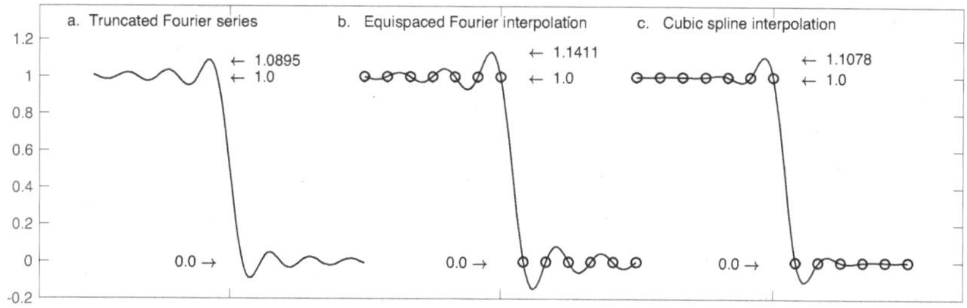

Figure 2.3 Three examples of the Gibbs phenomenon. The PS case in part (b) is generic for infinitely smooth interpolants near to the jump (such as for splines in their infinite order limit).

(a) Truncated Fourier series. The partial sums $\sum_{k=-N}^{N} \hat{u}_k e^{ikx}$ overshoot the jump at each side by about 9%. As $N \to \infty$, the region in which this occurs gets narrower, but the peak height converges to $\frac{1}{2} + \frac{1}{\pi} \int_0^{\pi} \frac{\sin \xi}{\xi} d\xi \approx 1.0895$. The error in the partial sum (in the maximum norm) will be $O(1)$ near the jump and $O(1/N)$ away from it.

(b) Equispaced Fourier interpolation. Another version of Gibbs phenomenon occurs if one performs equispaced interpolation with trigonometric functions. In this case, the overshoot approaches 14% of the height of the jump as the density of the interpolation points increases (or one interpolates over an increasingly wide interval).[4]

(c) Cubic spline interpolation. The peak value of the Gibbs overshoot is now $\frac{1}{6}\left(8 - 2\sqrt{2} - \sqrt{3} - \sqrt{6}\right) \approx 1.1078$, but the oscillations decay in this case exponentially fast[5] rather than only inversely proportionally to the distance from the jump.

In many cases, the Gibbs phenomenon is a serious source of error when applying high-order approximations to nonsmooth data. However, solving wave equations can sometimes be exceptional, as described in Fornberg (1996), Section 4.2, and (repeated) in Fornberg and Flyer (2015b), Section 2.1.3.2. When a step (or highly irregular) solution propagates sideways, oscillatory Gibbs-type errors can produce favorable numerical cancellations (also in variable coefficient cases), leaving a very accurate solution behind.

by J. Gibbs. In certain situations, such as when given a truncated Fourier series together with discontinuity locations, the Gibbs oscillations can be very accurately eliminated (Driscoll and Fornberg, 2001; Eckhoff, 1998; Gottlieb and Shu, 1997).

4 Polynomial interpolation features this same limit next to a jump.

5 With the ratio of successive amplitudes approaching $2 - \sqrt{3} \approx 0.2679$, just as is noted for cardinal splines in Section B.2.5. For analysis, it is often useful to view step data as the superposition of translates of cardinal data.

2.3 Some Possible Concerns about Equispaced PS Methods

PS discretizations are generally very well suited for approximating periodic problems on equispaced grids. Figures 2.1 and 2.2 give the impression that the periodic PS method then is optimal in that it handles all Fourier modes exactly. However, a couple of concerns can be raised. The relatively large FD weights also far from the point of approximation are unnatural for approximating strictly local operators (such as an integer order derivative). Above 1-D, an additional issue arises, as indicated in Figure 2.4. Suppose we wish to approximate the operator $\left\{\frac{\partial}{\partial x} + \frac{\partial}{\partial y}\right\}$ at the center point on the grid. The PS approximations for $\frac{\partial}{\partial x}$ and $\frac{\partial}{\partial y}$ will use weights at the nodes marked by squares and circles, after which these results are added. However, $\left\{\frac{\partial}{\partial x} + \frac{\partial}{\partial y}\right\}$ can alternatively be seen as a directional derivative in the 45° direction, shown by triangles. Since PS weights decay very slowly with distance, the approximations along the coordinate axes pick up lots of quite irrelevant distant information, while ignoring local information that is available in the direction that corresponds to the combined operator. Discussions in Fornberg et al. (2010) suggest that, even for an operator such as $\frac{\partial}{\partial x}$, mathematically completely determined by data just along the x-axis, there can be relevant information also off the axis, see Figure 2.5. In cases when some additional information is available about a multivariate function for which we wish to approximate $\frac{\partial}{\partial x}$, for example, that it satisfies a PDE such as Laplace or Helmholtz equation, that can be effectively utilized for obtaining compact approximations, as described in Section 4.3.4.

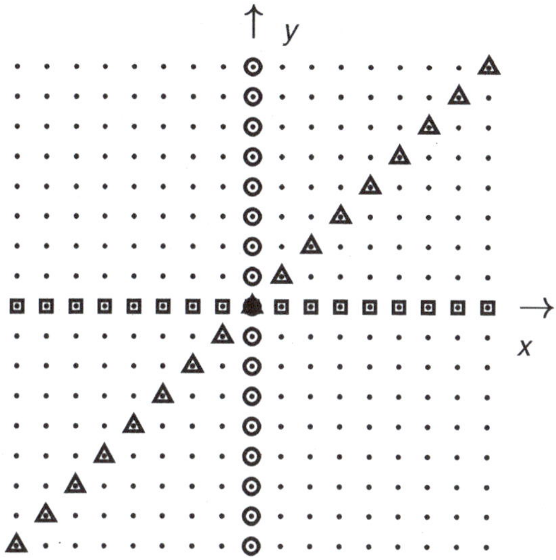

Figure 2.4 PS discretizations for $\frac{\partial}{\partial x}$ and $\frac{\partial}{\partial y}$ (marked by squares and circles, respectively) versus the directional derivative $\left\{\frac{\partial}{\partial x}+\frac{\partial}{\partial y}\right\}$.

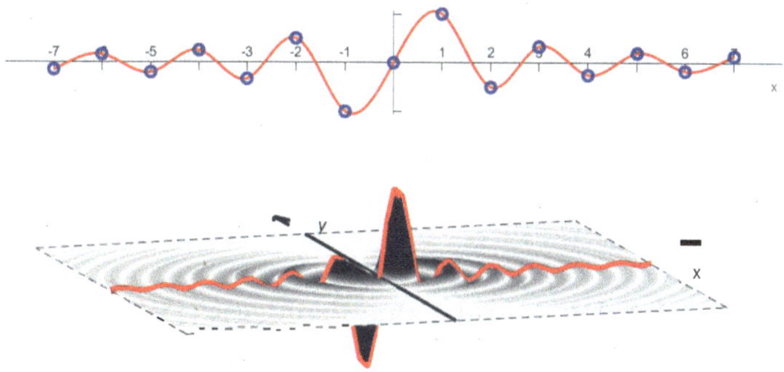

Figure 2.5 Illustration of weight patterns for approximating $\frac{\partial}{\partial x}$. Top: Standard PS approximation in 1-D, Bottom: A corresponding 2-D pattern heuristically arrived at in Fornberg et al. (2010) (displaying the function $\frac{x}{8}\,{}_0F_1(3,-\frac{1}{4}(x^2+y^2)) = \frac{x}{x^2+y^2}J_2(\sqrt{x^2+y^2})$).

3

FD Approximations for Ordinary Differential Equations

Linear algebra and ordinary differential equations (ODEs) are two areas where modern program packages and library routines have reduced the need for many users to deal with algorithmic technicalities. For ODEs, MATLAB/Chebfun users will find much useful information in Trefethen et al. (2017). However, there are also situations where packages do not apply efficiently, and some user customization is needed. Many textbooks on numerical analysis provide excellent introductions to the usage of FD approximations for both ODEs and partial differential equations (PDEs). For more in-depth discussions, see for example Ascher and Petzold (1988); Gustafsson (2008); Iserles (2012); LeVeque (2007) and, focusing entirely on ODEs, Ascher et al. (1995); Butcher (2016); Deuflhard and Bornemann (2002); Lambert (1991) and the two volumes Hairer et al. (1992); Hairer and Wanner (2002). The present treatment is brief, focusing on how the "basic" FD formulas, as derived in Chapter 1, enter perhaps more immediately than is always recognized into well-established numerical ODE procedures. A common theme, seen already in Chapter 2, will be the benefits obtained and considerations required for reaching high orders of accuracy.

3.1 ODE Initial Value Problems (IVP)

A general first-order ODE can be written in the form

$$\frac{dy}{dt} = f(t, y), \quad y(0) = y_0. \tag{3.1}$$

Numerically, most algorithms work similarly if $y(t)$ and $f(t, y(t))$ are scalar- or vector-valued functions, so our focus (for notational simplicity) will be on the former.[1] Higher order ODEs, say $y''' = f(t, y, y', y'')$, with initial conditions (ICs) on $y(0)$, $y'(0)$, and $y''(0)$ are readily rewritten as a first-order system following the variable changes $y_1(t) = y(t)$, $y_2(t) = y'(t)$, $y_3(t) = y''(t)$, and are usually handled that way numerically.[2] The most basic numerical scheme for solving (3.1) is known as the Forward Euler (FE) method, immediately obtained from (1.2)

$$y(t + k) = y(t) + k\, f(t, y(t)) + O(k^2), \tag{3.2}$$

where, for computing, the $O(k^2)$ term is dropped.[3] With this size error at each time step (of length k), FE is described as a first-order method.[4] This low accuracy order makes FE grossly inefficient in virtually all application contexts.

All four of the most important families of numerical ODE IVP methods have FE as their most basic member, and then enhance FE in different ways, as described in the next Sections 3.1.1– 3.1.4. These methods can all be seen as FD formulas in various amounts of disguise.

3.1.1 Linear Multistep (LM)

With the notation used in the FD stencil sketches in (1.3) and Section 1.2.1.5, we illustrate the FE scheme as

FE:
```
○
□ □ .
```

In this stencil sketch, the t-axis goes to the right; squares refer to y-values (here at t and $t+k$), and the circle to a y'-value (here at t). It is assumed that numerical values (for y and $y' = f(t, y)$) are known up through time t. With these values and the FD weights corresponding to the stencil, $y(t + k)$ is calculated, and the stencil is moved one step forward in time.

The weights in the main three LM methods can all be obtained immediately from the algorithm in Section 1.2.1.5. In Table 3.1, p denotes the order of accuracy. In the stencil shape sketches in the table, the total number of non-zero entries is $p + 2$. The AB schemes are *explicit* – the value $y(t + k)$ is

1 Runge–Kutta methods above order 4 can be exceptional in that their derivations depend on terms commuting in a way that can fail and cause a loss of accuracy for systems, see Butcher (2016), Section 316, and Lambert (1991), Section 5.8.

2 Many software environments assume this and do not provide algorithms specific for higher-order ODEs.

3 In contrast to Chapters 1 and 2, we now denote the independent variable by t rather than x, as in ODE contexts it often corresponds to time, and we denote the discretization step by k rather than by h.

4 Strict proofs for an $O(k^1)$ error bound after $O(1/k)$ time steps are given in many introductory texts.

Table 3.1 *The three main families of linear multistep methods. With p denoting the method's order of accuracy, the stencil weights can then obtained by choosing the shown values for s, d, n, m in the algorithm in Section 1.2.1.5.*

Abbrev-iation	Name	Stencil shape t-direction →	Parameters s	d	n	m
AB	Adams–Bashforth	○ … … ○ ○ □ □	$1-p$	$p-1$	1	1
AM	Adams–Moulton	○ … … ○ ○ ○ □ □	$2-p$	$p-1$	1	1
BD	Backward differentiation	○ □ … … □ □ □	p	0	p	1

obtained directly by known values for y and $y' = f$ at previous times t, $t-k, \ldots$. The AM and BD schemes are *implicit*, requiring at level $t+k$ the solution of a system combining the values of $y(t+k)$ and $f(t+k, y(t+k))$. Partly offsetting this disadvantage are their larger *stability domains* – see Section 3.2.3. *Predictor–corrector schemes* use an explicit scheme to predict a value at the next time level, and then use this in a scheme that otherwise would have been implicit, combining effective explicitness with improved stability domains. In the literature there are various derivations for the weights in the AB, AM, and BD schemes of a general order p. As indicated in the table, these schemes can all be seen more simply as quite standard FD approximations of the first derivative.[5] A minor inconvenience with LM methods is that some other method is needed for the first few time levels (such as using a few initial Runge–Kutta steps). This also makes LM methods awkward in combination with time splitting.[6]

3.1.2 Runge–Kutta Methods

There exist a vast number of RK methods, both explicit and implicit. Denoting the number of stages by s and the accuracy order by p, we can write Forward Euler (FE) as

$$s = 1,\ p = 1 \qquad \frac{d^{(1)} = k\, f(t, y(t))}{y(t+k) = y(t) + \left\{d^{(1)}\right\}} \qquad \begin{array}{c|c} 0 & \\ \hline & 1 \end{array}.$$

In this and the next few cases, the compact *Butcher diagram* for the scheme is shown to the right. The order of accuracy can be increased by adding more

[5] For the AB and AM schemes, one common derivation starts by rewriting $y' = f$ as $y(t+k) - y(t) = \int_t^{t+k} f\, dt$ and then replacing f by the Lagrange interpolant based on interpolated f-values.

[6] See Section 4.2.4.

stages, each taking the form of a single FE-like discretization, and with coefficients chosen to produce the desired overall accuracy. Examples of explicit $s = p = 2, 3, 4$ schemes are shown below:

$$
\begin{array}{ll}
s = 2, & d^{(1)} = k\, f(t, y(t)) \\
p = 2 & d^{(2)} = k\, f(t + \tfrac{1}{2}k, y(t) + \tfrac{1}{2}d^{(1)}) \\
\hline
 & y(t+k) = y(t) + \{d^{(2)}\}
\end{array}
\qquad
\begin{array}{c|cc}
0 & & \\
1/2 & 1/2 & \\
\hline
 & 0 & 1
\end{array}
$$

$$
\begin{array}{ll}
s = 3, & d^{(1)} = k\, f(t, y(t)) \\
p = 3 & d^{(2)} = k\, f(t + \tfrac{1}{3}k, y(t) + \tfrac{1}{3}d^{(1)}) \\
 & d^{(3)} = k\, f(t + \tfrac{2}{3}k, y(t) + \tfrac{2}{3}d^{(2)}) \\
\hline
 & y(t+k) = y(t) + \{\tfrac{1}{4}d^{(1)} + \tfrac{3}{4}d^{(3)}\}
\end{array}
\qquad
\begin{array}{c|ccc}
0 & & & \\
1/3 & 1/3 & & \\
2/3 & 0 & 2/3 & \\
\hline
 & 1/4 & 0 & 3/4
\end{array}
$$

$$
\begin{array}{ll}
s = 4, & d^{(1)} = k\, f(t, y(t)) \\
p = 4 & d^{(2)} = k\, f(t + \tfrac{1}{2}k, y(t) + \tfrac{1}{2}d^{(1)}) \\
 & d^{(3)} = k\, f(t + \tfrac{1}{2}k, y(t) + \tfrac{1}{2}d^{(2)}) \\
 & d^{(4)} = k\, f(t + \phantom{\tfrac{1}{2}}k, y(t) + d^{(3)}) \\
\hline
 & y(t+k) = y(t) + \tfrac{1}{6}\{d^{(1)} + 2d^{(2)} + 2d^{(3)} + d^{(4)}\}
\end{array}
\qquad
\begin{array}{c|cccc}
0 & & & & \\
1/2 & 1/2 & & & \\
1/2 & 0 & 1/2 & & \\
1 & 0 & 0 & 1 & \\
\hline
 & 1/6 & 1/3 & 1/3 & 1/6
\end{array}
$$

This last scheme is traditionally the best-known RK scheme (often denoted RK4). Over a century after its discovery, it remains one of the most popular ODE IVP solvers available.[7] It transpires that, when seeking still higher orders of accuracy, s will need to be larger than p. Even so, schemes up though p around 8 are widely used (with $p = 8$ requiring $s = 11$ stages).

As for LM methods, implicit RK (IRK) methods can offer attractive accuracy and stability features (in exchange for having to solve a system of equations at each step). A well-known implicit $s = 2$, $p = 4$ scheme (Hammer–Hollingsworth) is given by

$$
\begin{array}{l}
s = 2,\ p = 4 \\
d^{(1)} = k\, f(t + (\tfrac{1}{2} - \tfrac{\sqrt{3}}{6})k, y(t) + \tfrac{1}{4}d^{(1)} + (\tfrac{1}{4} - \tfrac{\sqrt{3}}{6})d^{(2)}) \\
d^{(2)} = k\, f(t + (\tfrac{1}{2} + \tfrac{\sqrt{3}}{6})k, y(t) + (\tfrac{1}{4} + \tfrac{\sqrt{3}}{6})d^{(1)} + \tfrac{1}{4}d^{(2)}) \\
\hline
y(t+k) = y(t) + \tfrac{1}{2}\{d^{(1)} + d^{(2)}\}
\end{array}
\qquad
\begin{array}{c|cc}
\tfrac{1}{2} - \tfrac{\sqrt{3}}{6} & \tfrac{1}{4} & \tfrac{1}{4} - \tfrac{\sqrt{3}}{6} \\
\tfrac{1}{2} + \tfrac{\sqrt{3}}{6} & \tfrac{1}{4} + \tfrac{\sqrt{3}}{6} & \tfrac{1}{4} \\
\hline
 & \tfrac{1}{2} & \tfrac{1}{2}
\end{array}
$$

A vast number of both (explicit) RK and IRK variations are available with different trade-offs, as discussed in many of the ODE texts cited earlier. The opportunities to utilize parallel processing (multiple computational cores) within RK methods are quite limited (Iserles and Nørsett, 1990; Ketcheson and Bin-Waheed, 2014). However, RK methods can be used as building blocks within certain parallel-in-time frameworks, such as *Parareal* (Gander, 2015).

[7] The history of RK developments is summarized in Butcher (1996).

3.1.3 Taylor Series Methods

The idea is straightforward – include many more Taylor expansion terms in (3.2). For order p:

$$\underbrace{y(t+k) = y(t) + k\, f(t, y(t))}_{\text{FE}} + \frac{k^2}{2!}\frac{d}{dt} f(t, y(t)) + \cdots$$
$$+ \frac{k^p}{p!}\frac{d^{p-1}}{dt^{p-1}} f(t, y(t)) + O(k^{p+1}). \qquad (3.3)$$

Notable features of this approach include:

i. The order of accuracy (p) can often be made extremely high (say $p = 50$, or so).
ii. Once a long Taylor expansion is numerically available, it can be converted to continued fraction or Padé rational form (as done in Willers (1974) and (Fornberg and Weideman, 2011), respectively[8]) for improved accuracy, especially if integrating near to singularities. For Taylor-to-Padé conversion, see Section E.3.
iii. The stability domains for the Taylor series method increase with p (rather than severely decrease with p, as is the case with LM methods), see Section 3.2.3.

Regarding (i): The perhaps most "obvious" implementation (obtain the derivatives in (3.3) by repeated analytic differentiation of the governing ODE) is in most cases hopelessly inefficient, since the complexity increases rapidly:

$$y' = f$$
$$y'' = f\frac{\partial f}{\partial y} + \frac{\partial f}{\partial t}$$
$$y''' = f^2\frac{\partial^2 f}{\partial y^2} + f\left\{\left(\frac{\partial f}{\partial y}\right)^2 + 2\frac{\partial^2 f}{\partial t \partial y}\right\} + \left\{\frac{\partial^2 f}{\partial t^2} + \frac{\partial f}{\partial t}\frac{\partial f}{\partial y}\right\}$$
$$y'''' = f^3\frac{\partial^3 f}{\partial y^3} + f^2\left\{3\frac{\partial^3 f}{\partial t \partial y^2} + 4\frac{\partial f}{\partial y}\frac{\partial^2 f}{\partial y^2}\right\}$$
$$+ f\left\{\left(\frac{\partial f}{\partial y}\right)^3 + 5\frac{\partial^2 f}{\partial t \partial y}\frac{\partial f}{\partial y} + 3\frac{\partial^3 f}{\partial t^2 \partial y} + 3\frac{\partial f}{\partial t}\frac{\partial^2 f}{\partial y^2}\right\}$$
$$+ \left\{\frac{\partial^3 f}{\partial t^3} + 3\frac{\partial f}{\partial t}\frac{\partial^2 f}{\partial t \partial y} + \left(\frac{\partial f}{\partial y}\right)^2\frac{\partial f}{\partial t}\right\}$$
$$\cdots \quad \cdots$$

[8] In combination with following branching integration paths in the complex plane, the latter version is at present the computationally most effective approach available for solving the Painlevé equations; see also Fasondini et al. (2017) and Fornberg and Piret (2020), §11.5.

It is much better to start by considering the ODE (3.1) in the form

$$\frac{dy(t+k)}{dk} = f(t+k, y(t+k)). \tag{3.4}$$

Each time a truncated expansion[9]

$$y(t+k) = y(t) + a_1 k + a_2 k^2 + \cdots + a_n k^n \tag{3.5}$$

is substituted into the RHS of (3.4), reexpanded in powers of k, and integrated with respect to k, one more correct coefficient is gained in (3.5).[10]

Example 3.1.1 Compare for the parameter value $\mu = 5$ the convergence rates of some of the one-step ODE schemes above for solving the Van der Pol oscillator equation

$$\frac{d^2y}{dt^2} = \mu(1-y^2)\frac{dy}{dt} - y, \tag{3.6}$$

with initial conditions $y(0) = 1, y'(0) = 0$.

Solution Since (3.6) is a second-order ODE, it can be convenient to rewrite it as a first-order system

$$\begin{cases} \frac{dy_1}{dt} = y_2 & , \quad y_1(0) = 1, \\ \frac{dy_2}{dt} = \mu(1-y_1^2)y_2 & , \quad y_2(0) = 0. \end{cases} \tag{3.7}$$

Figure 3.1 shows the solution (with $y(t)$ and $y'(t)$ corresponding to $y_1(t)$ and $y_2(t)$) over the time interval $0 \le t \le 30$. It approaches a periodic state, with a period that increases with the value of the parameter μ. Since the solution features sharp transition zones, this example is a good candidate for numerical methods that use adaptively changing time steps. In spite of FE, RK4 and the Taylor series method all being one-step methods (and thus easily allowing for variable time steps), we use here fixed time steps k to simplify the comparison shown in Figure 3.2.[11] This figure shows convergence rates

[9] With $y(t)$ known at first from the ODE's IC and then from the result of the previous time step.

[10] Earlier coefficients are unchanged and need not be recomputed. A requirement for the Taylor series method to be applicable is that the ODE's RHS is of a relatively simple form, analytic in t and y. When this is the case, implementation requires only numerical operations on expansion coefficients, involving no symbolic computations. Key operations such as series addition, subtraction, multiplication, division, integration, and so on (aiming for just one more coefficient) can all be performed as one-line statements in MATLAB, and are conveniently written as *anonymous functions*; see Algorithm 3.1. For order of accuracy p, the cost per time step typically scales as $O(p^2)$. For a historical perspective, see Hairer et al. (1992), page 47. Requiring considerably more extensive software and computations, "automatic differentiation" can also generate leading terms of Taylor expansions (Barton et al., 1971; Chang and Corliss, 1994; Griewank and Walther, 2008; Neidinger, 2010).

[11] For (3.6), the loop over the time steps requires in neither of the FE, RK4, and the Taylor series method (arbitrary accuracy order) cases more than about 10 lines of MATLAB code.

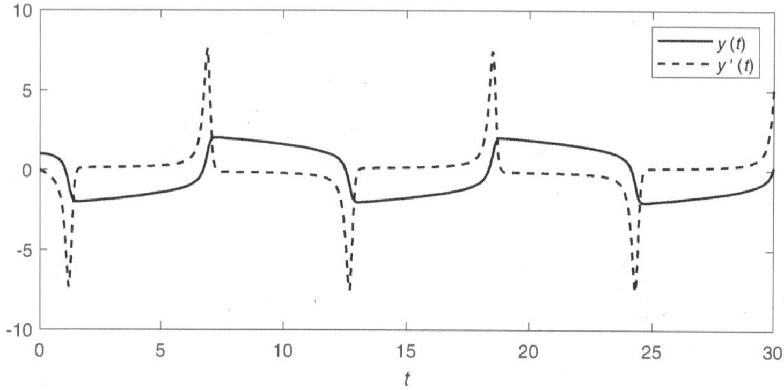

Figure 3.1 The solution over $0 \le t \le 30$ for the ODE in Example 3.1.1.

that match the respective method's order of accuracy. We can note: (i) FE is extremely ineffective computationally, and (ii) the effectiveness of very high-order methods is not adversely affected by the spikes in the solution. □

Example 3.1.2 Give a schematic MATLAB code section for carrying out a time step using the Taylor method on Example 3.1.1.

Solution The code section shown as Algorithm 3.1 is specific to (3.7), but generalizes readily to other ODEs or ODE systems. The first two of the remarks below relate to stepping in the complex plane rather than (as here) along a real t-axis:

i. The use of .' rather than just ' in the anonymous functions mult and div ensures the correct results also in the complex-valued case.
ii. Once local Taylor expansions have been obtained, these can very rapidly be evaluated for numerous different k-values, speeding up computations that extend over 2-D complex plane regions (rather than along 1-D curves or straight lines).
iii. The anonymous function div is not utilized in this Van der Pol example (since (3.7) contains no divisions), but is included as general reference. □

Many of the special functions in applied mathematics are given by linear ODEs with coefficients that are low-degree polynomials in the independent variable, which may be complex and we thus denote by z. In such cases, the Taylor series method can often be implemented to accuracy order p at a cost of only $O(p)$ rather than $O(p^2)$ operations per time step.

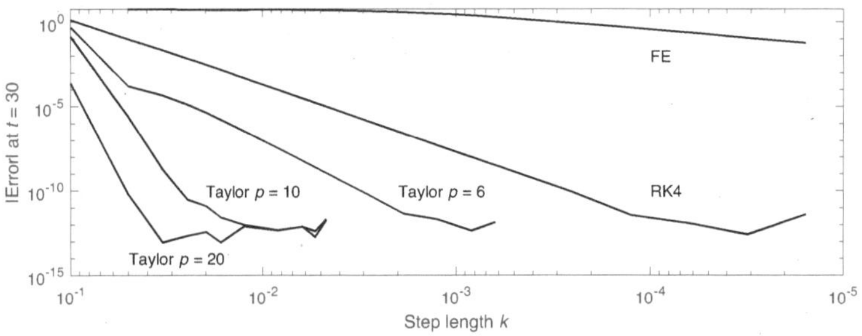

Figure 3.2 Numerical errors at final time $t = 30$, measured as the sum of the magnitude of the errors in $y(30)$ and $y'(30)$ (i.e., at a time where the solution is particularly rapidly changing, see Figure 3.1). All calculations were done in regular double precision. Higher orders of accuracy are usually not only more cost-effective but also (as seen previously in Figure 1.2) less affected by rounding errors.

Example 3.1.3 Evaluate numerically the Gauss hypergeometric function

$$
\begin{aligned}
{}_2F_1(a,b;c;z) = 1 &+ \frac{a \cdot b}{1! \cdot c} z + \frac{a(a+1) \cdot b(b+1)}{2! \cdot c(c+1)} z^2 \\
&+ \frac{a(a+1)(a+2) \cdot b(b+1)(b+2)}{3! \cdot c(c+1)(c+2)} z^3 + \cdots \qquad (3.8)
\end{aligned}
$$

at an arbitrary complex z-value.[12] These references discuss also conditioning issues that can arise if $|a|$, $|b|$, and/or $|c|$ are very large and certain a, b, c-combinations that give rise to singularities.

Solution The governing ODE for $y(z) = {}_2F_1(a,b;c;z)$ can be written as

$$(1-z)\, z\, y''(z) + (c - (a+b+1)\, z)\, y'(z) - a\, b\, y(z) = 0, \qquad (3.9)$$

where we are interested only in the solution that satisfies $y(0) = 1$. It follows from (3.9) that the only singularity of ${}_2F_1(a,b;c;z)$ in the finite complex plane is a branch point at $z = 1$. From this (and also directly from (3.8)) it follows that the radius of convergence of the Taylor expansion around the origin is $R = 1$. From the functional identities (15.8.2) and (15.8.5) in Olver et al. (2010) (mapping $z \to \frac{1}{z}$ and $z \to 1 - \frac{1}{z}$) it follows further that evaluations of ${}_2F_1(a,b;c;z)$ for any z can be carried out by either one of two evaluations within the complex plane region $|z| \le 1 \cap \mathrm{Re}(z) \le \frac{1}{2}$, shown in Figure 3.3. The Taylor series method can then be implemented effectively as follows:

- For z-values within the inner disk, evaluate directly using (3.8).

[12] Several computational options (not including the present Taylor series approach) are compared in Pearson (2009); Pearson et al. (2017).

Algorithm 3.1 Schematic MATLAB algorithm considered in Example 3.1.2.

```
% Anonymous functions for series multiplication and division.
% Given entries 1,2,...,n in a and b and for division also
% entries 1,2,...,n-1 in c (row vectors), create
% the nth entry of the product or quotient, respectively
mult = @(a,b,n) a(1:n)*b(n:-1:1).';                      % c = a*b

div  = @(a,b,c,n) (a(n)-b(2:n)*c(n-1:-1:1).')/b(1);    % c = a/b

% Initialize parameters
p = ....;              % Set accuracy order
k = ....;              % Set the size of the time step

% Allocate space for vectors
y1 = zeros(1,p+1);  % Vectors to hold the Taylor expansions for
y2 = zeros(1,p+1);  % y1 and y2 during each time step

tv = zeros(1,p);    % Vector to hold temporary expansion for (1-y1^2)

% Set start values for y1 and y2 for the step of size k
% y1(1) = ....; y2(1) = ....;


% Carry out one time step of size k; Repeat for multiple steps
for n = 1:p                              % Build up p+1 terms in the
    tv(n)   = -mult(y1,y1,n);            % expansions for y1 and y2
    if n==1; tv(1) = tv(1)+1; end        % Add one more term to the
    t1      = mu*mult(y2,tv,n)-y1(n);  % 1-y^2 expansion that is held in
    y1(n+1) = y2 (n)/n;                  % tv. Find new last term for the
    y2(n+1) = t1 /n;                     % full ODE. Integrate to gain
end                                      % one more term for y1. Integrate
y1(1) = polyval(y1(p+1:-1:1),k);         % to gain one more term for y2
y2(1) = polyval(y2(p+1:-1:1),k);         % Values for y1 and y2 after the
                                         % time step
```

- For z-values between this circle and the outer boundary (e.g., at the point illustrated by the small square), evaluate at the inner circle edge (shown by small circle) and step to the evaluation point z using the Taylor series method implemented to a high order of accuracy.

One finds, after substituting

$$\begin{array}{ccccccccccc} y(z_0+k) & = & s_1 & + & s_2 k & + & s_3 k^2 & + & s_4 k^3 & + & \cdots \\ & & \updownarrow & & \updownarrow & & & & & & \\ & & y(z_0) & & y'(z_0) & & & & & & \end{array} \tag{3.10}$$

into

$$(1-(z_0+k))(z_0+k)\frac{d^2y(z_0+k)}{d^2k}+(c-(a+b+1))(z_0+k)\frac{dy(z_0+k)}{dk} \\ -a\,b\,y(z_0+k)=0$$

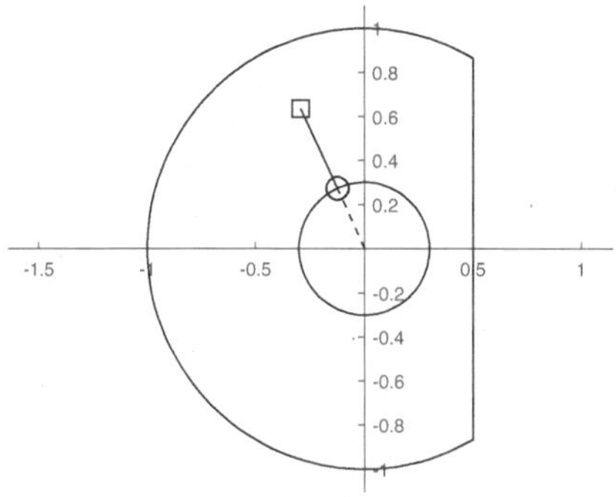

Figure 3.3 Geometry for the Taylor series approach for evaluating ${}_2F_1(a, b; c; z)$. The outer boundary encloses $|z| \le 1 \cap \mathrm{Re}(z) \le \frac{1}{2}$. The inner circle is here shown with radius 0.3.

Algorithm 3.2 Coefficient recursion for evaluating the ${}_2F_1(a, b; c; z)$ function.

```
a = ...; b = ...; c = ...; % Set 2F1 parameters
z = ...; p = ...;          % Set expansion point and order of expansion
s(1) = ...;  s(2) = ...;   % Give the first two Taylor terms
d1 = -a*b; d2 = 1-a-b; d3 = c-(1+a+b)*z; d4 = d3; d5 = 2-4*z;
d6 = 2*z*(z-1); d7 = d6; d8 = d6;
for n = 3:p
    s(n) = (d1*s(n-2)+d3*s(n-1))/d6; % Compute next s-coefficient
d2 = d2-2; d1 = d1+d2; d4 = d4+d5; d3 = d3+d4; d7 = d7+d8; d6 = d6+d7;
end
```

(having followed the idea from (3.4)) and equating coefficients for k, a 3-term recursion relation between the coefficients s_n in (3.10). Given values for $y(z_0)$ and $y'(z_0)$, that is, for s_1 and s_2, we can, by these relations, explicitly calculate any number of further Taylor coefficients $s_3, s_4, s_5, \ldots$. From these, we can then evaluate $y(z_0 + k)$ and $y'(z_0 + k)$, that is, obtain the information needed for the next Taylor step (two start values, since the ODE is of second order). After some algebraic simplifications, the code for finding $s_3, s_4, s_5, \ldots, s_p$ can be written as sketched out in Algorithm 3.2. □

One can note a conceptual similarity between the Taylor series method, as used here, and the circle–chain analytic continuation approach described, for example, in Fornberg and Piret (2020), Section 3.2.1. While the latter is hopelessly ill conditioned for numerical work (Trefethen, 2023), the Taylor series approach is entirely well conditioned, since it utilizes the governing ODE at each continuation step (rather than only analyticity).

3.1.4 Extrapolation Methods

The Gragg–Bulirsch–Stoer (GBS) method is a particularly effective example of an extrapolation method (Gragg, 1965; Hairer et al., 1992; Lambert, 1991). Like LM methods, it is very closely connected to regular FD formulas. With N even, it advances $y' = f(t, y)$ from time t to time $t + Nk$ by the following sequence of steps:

First step	$y(t+k) = y(t) + kf(t, y(t))$	Forward Euler (FE)
Then N steps, $n = 1, 2, \ldots, N$	$y(t+(n+1)k) = y(t+(n-1)k)$ $+2kf(t+nk, y(t+nk))$	Centered FD2
Accepted value at $t + Nk$	$y^*(t+Nk) = \frac{1}{4}\big(y(t+(N+1)k) + 2y(t+Nk)$ $+y(t+(N-1)k)\big)$	Average

Surprisingly, it transpires that the error after this sequence of steps possesses an expansion

$$y^*(t+Nk) - \text{Exact} = c_1k^2 + c_2k^4 + c_3k^6 + \cdots$$

that contains only even powers of k. This makes Richardson extrapolation (Section E.1) particularly effective. In certain contexts, such as when time stepping wave equations, the GBS approach provides an opportunity for increasing computational efficiency by means of using multiple computational cores in parallel for the separate k-values, see Section F.4.

3.2 Some Key Concepts Related to Convergence

3.2.1 Convergence

Two conditions suffice for *Convergence* to occur (of the numerical to the analytical solution, as the step size k goes to zero, ignoring rounding errors): *Consistency* and *Zero-stability*. Accuracy order $p = 1$ is more than is needed for consistency, so we need not consider that condition for any computationally relevant scheme. Once zero-stability (and thereby convergence) is assured, the rate of convergence will be given by the scheme's formal order of accuracy.

3.2.2 Zero-stability and Root Condition

The test for *zero-stability* (the root condition) is straightforward: Apply the scheme to the trivial ODE $y'(t) = 0$. If the scheme then has no growing solutions, zero-stability is satisfied.

Example 3.2.1 Consider the LM scheme

$$y(t+k) = \{3y(t) - 2y(t-k)\} + k\left\{\frac{1}{2}y'(t) - \frac{3}{2}y'(t-k)\right\}.$$

Determine its order of accuracy and whether it is convergent.

Solution For LM methods, the order of accuracy can be determined by successively substituting $y(t) = 1, t, t^2, t^3, \ldots$ and recording the first case when the scheme is not exactly satisfied.[13] In the present case, this occurs for $y(t) = t^3$, giving a residual $\frac{7}{2}k^3 \neq 0$. The error per step is therefore $O(k^3)$, and the scheme is second-order accurate.

Applied to $y'(t) = 0$, we obtain $y(t+k) = \{3y(t) - 2y(t-k)\}$. This is a linear recursion relation with characteristic equation $r^2 - 3r + 2 = 0$ and roots $r_1 = 1$, $r_2 = 2$. The recursion relation has as its general solution $y(t+nk) = c_1 \cdot 1^n + c_2 \cdot 2^n$. With $n = O(1/k)$, this grows without bound.[14] The scheme fails the condition for zero-stability and is thus not convergent as $k \to 0$. □

For the AB, AM, and RK schemes of order one or higher, the characteristic equation becomes $r - 1 = 0$, and these schemes are therefore all convergent to the ODE solution as $k \to 0$. The situation for the BD schemes is more complicated. The polynomial in r to test for its roots will have as its coefficients the entries on the corresponding row in Table 1.3. Figure 3.4 shows these roots for orders 1–8. These satisfy the condition for zero-stability only up through order $p = 6$.

We also note that a scheme that is at least first-order accurate will be exact for the $y(t) \equiv 1$ solution to $y'(t) = 0$, implying that the root condition polynomial will always have $r = 1$ (exactly) as a root. When changing the ODE from the trivial $y'(t) = 0$ to $y'(t) = f(t, y(t))$, one can heuristically think of all the roots becoming perturbed by $O(k)$. Roots inside the unit circle will (as $k \to 0$) remain spurious and correspond to modes that vanish during time stepping. The root at $r = 1$, perturbed by $O(k)$, will alone represent the full solution to the ODE.[15]

[13] RK methods require for determining orders of accuracy a more elaborate procedure involving Taylor expansions.

[14] Because of machine rounding errors, one must assume c_1 and c_2 to be nonzero.

[15] In the simple case of $y'(t) = \lambda y(t)$ approximated by FE, this is seen explicitly: $y(t+k) = (1 + \lambda k)y(t)$ implies $y(t) = y(0)\lim_{k\to 0}(1 + \lambda k)^{t/k} = y(0)\, e^{\lambda t}$.

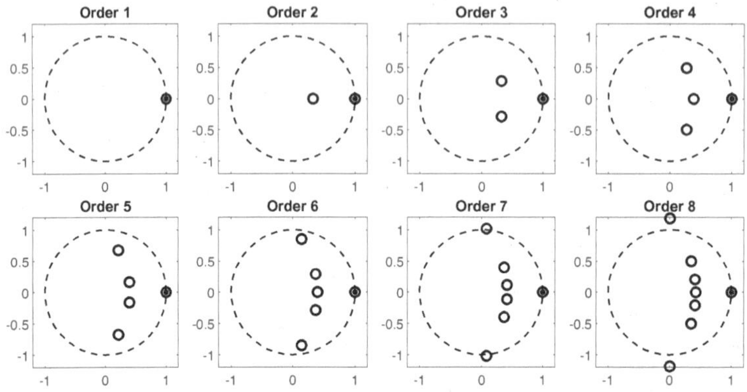

Figure 3.4 The roots of the characteristic polynomials for BD methods up through order $p = 8$. The unit circle is dashed; the root at $r = 1$ is shown with a central dot.

3.2.3 Absolute Stability and Stability Domain

Loosely speaking, these quantities characterize how large the step size k can be chosen without encountering artificially growing solutions. This is of key importance primarily in two contexts:

i. when solving *stiff* systems of ODEs (e.g., systems with some components decaying at a much faster rate than others),
ii. understanding convergence versus divergence for FD approximations to time-dependent PDEs (cf. Section 4.1.3.2).

Example 3.2.2 Consider FE applied to solving $y' = -10y$, $y(0) = 1$. What step sizes k can be used?

Solution The FE approximation becomes $y(t + k) = y(t) - 10ky(t) = (1 - 10k)y(t)$. While the analytic solution $y(x) = e^{-10t}$ decays toward zero, the FE approximation does this only when $|1 - 10k| < 1$, that is, the scheme is *absolutely stable* for $k < 1/5$. □

Stability domains (regions of absolute stability) provide a convenient approach for obtaining similar bounds on k for general ODEs and solvers. In this respect, linearization preserves the essential feature of the ODE, and it suffices to apply the solver to the test problem $y' = \lambda y$ (where λ may be complex). For each ODE solver, one can then plot a domain in a complex $\xi = \lambda k$ plane

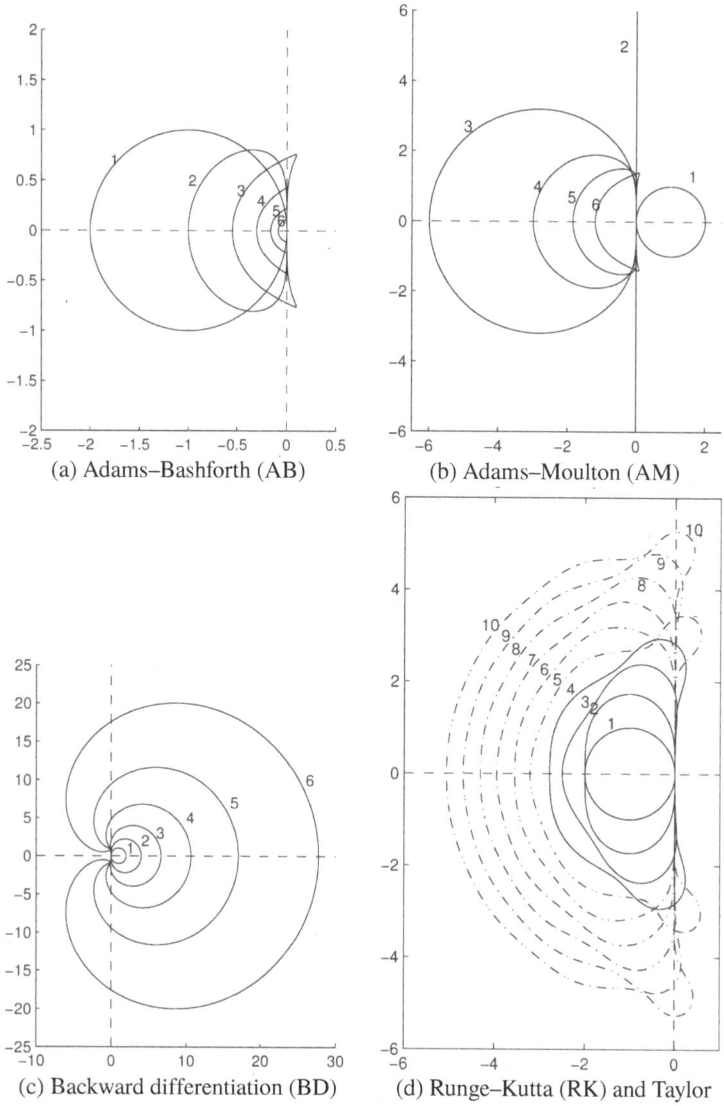

Figure 3.5 Boundaries of the stability domains for the five classes of ODE solvers. Note: (i) The different axes scales in the subplots (however, in all cases the same along the two axes), and (ii) The stable regions (ξ in the complex plane) are *inside* the shown curves in parts (a) and (d) and *outside* the curves in part (c). For part (b), stability is outside the circle for $p = 1$, to the left of the imaginary axis for $p = 2$, and inside the remaining curves. Figure adapted from Fornberg and Flyer (2015b).

such that the solver, applied to this test problem, has no growing solutions. Figure 3.5 illustrates the domains for the ODE solvers discussed above.[16] For all IRK $s = m, p = 2m,\ m = 1, 2, 3, \ldots$ schemes (with the Hammer–Hollingsworth example corresponding to $m = 2$), the stability domain is exactly the complex left half-plane, perfectly matching where the ODE $y' = \lambda y$ does not have growing solutions. Both for ODEs and even more when solving time-dependent PDEs, these domains tell much about which solvers are effective in different situations.

Returning to Example 3.2.2, the FE scheme is the same as AB1, and the ODE corresponded to $\lambda = -10$. The intersection of the $p = 1$ curve in Figure 3.5(a) with the negative ξ-axis occurs at $\xi = \lambda k = -2$. With $\lambda = -10$, this again gives $k < 1/5$.

Example 3.2.3 Determine the stability domain for the AB2 scheme

$$y(t+k) = y(t) + k\left\{\frac{3}{2}y'(t) - \frac{1}{2}y'(t-k)\right\}.$$

Solution Applying the scheme to $y' = \lambda y$ gives the 3-term linear recursion relation $y(t+k) - (1 - \frac{3}{2}\xi)y(t) + \frac{1}{2}\xi y(t-k) = 0$, with characteristic equation $r^2 - (1 - \frac{3}{2}\xi)r + \frac{1}{2}\xi = 0$. We want to plot the region in the ξ-plane for which both its roots are inside or on the unit circle. Although it might be tempting to approach this task by first solving this quadratic equation for its roots r_1 and r_2, it is much better to instead rewrite it as $\xi = \frac{2r(r-1)}{3r-1}$. The edge of the stability domain can then be obtained by setting $r = e^{i\theta}$ and plotting ξ for $0 \le \theta \le 2\pi$, giving the curve marked "2" in Figure 3.5(a). □

Example 3.2.4 Calculate the curves shown in Figure 3.5(d).

Solution Applying the Taylor schemes to $y' = \lambda y$ gives, for order p, the relation

$$r = 1 + \frac{\xi}{1!} + \frac{\xi^2}{2!} + \cdots + \frac{\xi^p}{p!}. \tag{3.11}$$

For explicit s-stage, order-p RK schemes with $s = p$, it transpires (i) as noted above, that such schemes are possible only for $s = p = 1, 2, 3, 4$ and (ii) that while these schemes can have several free parameters, their stability domains do not depend on these, and are also given by (3.11). One numerical procedure

[16] When time stepping wave equations, it is important that the stability domains contain a section of the imaginary axis around the origin, see Section 4.1.3.2. This is the case for AB methods only with accuracy orders $\{3, 4\}$, $\{7, 8\}$, $\{11, 12\}, \ldots$ and AM methods of orders $\{1, 2\}$, $\{5, 6\}$, $\{9, 10\}, \ldots$ (shown and generalized to predictor–corrector methods in Ghrist et al. (2015); tuning of parameters in AB–AM-type methods for improved stability domains is considered in Krogh (1966)). Details near the origin in the BD case are shown in Fornberg (1996), Figure G.2-4 and Fornberg and Flyer (2015b), Figure 1.3.

for computing the domains is obtained by first noting that $r = 1$ corresponds to $\xi = 0$. Then r can be incremented in small steps around the unit circle and, after each increment, Newton's method can be used to obtain the ξ-value closest to the previous iterate.[17] □

3.2.4 The Dahlquist Stability Barriers

The linear multistep (LM) schemes illustrated in Table 3.1 extend backward in time in either y or in $y' = f$, but not in both. For accuracy order p, each scheme contains exactly $p + 2$ nonzero coefficients. For a scheme of order p, it would seem more logical to extend backward about $p/2$ steps in both variables, as such a stencil would be more compact, rely less on further away information and, for the same order of accuracy, have a much smaller error constant. Attempting that would, however, run afoul of the first of the two *Dahlquist stability barriers.*[18] These are as follows:

First barrier: The accuracy order p for a zero-stable q-level LM scheme is limited to $p \leq q$ if q is even and to $p \leq q + 1$ if q is odd. If the scheme is explicit, then $p \leq q - 1$.

Second barrier: No explicit LM scheme can be A-stable, that is, have a stability domain that includes the entire LHP (left half-plane).

Example 3.2.5 We note in Table 3.1 that the AB scheme of order p uses a total of $p + 1$ levels, but only two of these for y. Since the orders of accuracy in general can be increased by one for every additional entry that is used, the idea raised above can also be expressed as attempting to increase the accuracy by also including more back levels for y (similarly to how many of these are used in the BD schemes). Is that practical?

Solution With a total of $q = p + 1$ levels in an AB scheme of order p, the first Dahlquist barrier tells that the accuracy order of these schemes already is as high as possibly can be achieved without losing zero stability. Attempting what is suggested would by necessity lead to a divergent scheme. □

In the more than 60 years since their introduction, numerous additional variations and extensions of the stability barriers have been discovered, as surveyed in Givoli (2023).

17 Another option is to make a contour plot of $|r(\xi)|$ with one as the only contour level. This method may also show disconnected "islands" of stability (typically of no computational significance).

18 Introduced in Dahlquist (1956) and Dahlquist (1963), respectively.

3.3 ODE Boundary Value Problems (BVP)

In BVP, the supplementary information to obtain a unique solution is given as boundary conditions (BC) at two (or more) locations of the independent variable. While we denoted this variable by "t" in the context of IVP, it is more often a space variable in BVP contexts, which we denote by "x." As model problems, we focus on the linear ODE

$$y'' = p(x)y' + q(x)y + r(x), \tag{3.12}$$

and the nonlinear ODE

$$y'' = f(x, y, y'), \tag{3.13}$$

in both cases with BC of the form: $y(a) = \alpha, y(b) = \beta$. Many introductory numerical textbooks have at least a chapter on ODE BVPs, for example, Bradie (2006); Burden and Faires (2015); Cheney and Kincaid (2012); Süli and Mayers (2003), presenting a number of approaches:

i. if linear and of second order, the BVP can be solved as a combination of two IVP solutions;
ii. if nonlinear; "shooting" (vary an additional IC for y' to find a solution that matches the BC at the opposite end) combined with Newton or secant root finding generally converges rapidly;
iii. directly approximate the ODE and BC with FD;
iv. variational (Ritz) methods;
v. Galerkin methods;
vi. Chebyshev spectral methods.

Given the theme of this book, the focus below will be on (iii), and especially on some options for increasing the order of accuracy beyond "basic" second order.

3.3.1 Second-Order Approximations

3.3.1.1 Linear Case

With equispaced nodes $x_0, x_1, \ldots, x_N$ over $[a, b]$, the derivatives in (3.12) are most simply approximated by

$$y'_n = \frac{1}{2h}(y_{n+1} - y_{n-1})$$

and

$$y''_n = \frac{1}{h^2}(y_{n+1} - 2y_n + y_{n-1}),$$

leading to a linear system of the form

$$
\left[\begin{array}{ccccccc}
1 & & & & & & \\
\hline
\times & \times & \times & & & & \\
& \times & \times & \times & & & \\
& \ddots & \ddots & \ddots & & & \\
& & & \times & \times & \times & \\
& & & & \times & \times & \times \\
\hline
& & & & & & 1
\end{array}\right]
\left[\begin{array}{c} y_0 \\ y_1 \\ y_2 \\ \vdots \\ \vdots \\ y_{N-1} \\ y_N \end{array}\right]
=
\left[\begin{array}{c} \alpha \\ \hline r(x_1) \\ r(x_2) \\ \vdots \\ \vdots \\ r(x_{N-1}) \\ \hline \beta \end{array}\right]
\begin{array}{c} \text{Left BC} \\ \\ \text{ODE at} \\ \text{interior} \\ \text{points} \\ \\ \text{Right BC} \end{array},
\tag{3.14}
$$

where the elements on row n are

$$
\frac{1}{h^2}\left[\begin{array}{ccc} 1 & -2 & 1 \end{array}\right] - p(x_n)\frac{1}{2h}\left[\begin{array}{ccc} -1 & 0 & 1 \end{array}\right] - q(x_n)\left[\begin{array}{ccc} 0 & 1 & 0 \end{array}\right].
$$

The ODE BVP (3.12) is assured to have a unique solution if $q(x) > 0$. This will also be true for (3.14) if h is sufficiently small (as seen by Gershgorin's theorem).

Derivative BCs are usually implemented by adding a "ghost point" outside the boundary, then centering both a BC approximation and an ODE approximation at the boundary point, followed by eliminating the ghost point weight.

3.3.1.2 Nonlinear Case

FD discretization of (3.13) leads to a system of nonlinear ODEs of a form well suited to Newton iterations.

Newton's method for a general nonlinear system: Given a system

$$
\left\{\begin{array}{lcl}
f_1(x_1, x_2, \ldots, x_n) & = & 0 \\
f_2(x_1, x_2, \ldots, x_n) & = & 0 \\
\vdots & \vdots & \vdots \\
f_n(x_1, x_2, \ldots, x_n) & = & 0,
\end{array}\right.
\tag{3.15}
$$

the Newton updates to an approximation $x_1, x_2, \ldots, x_n$ are obtained by solving

$$\underbrace{\begin{bmatrix} \frac{\partial f_1}{\partial x_1} & \frac{\partial f_1}{\partial x_2} & \cdots & \frac{\partial f_1}{\partial x_n} \\ \frac{\partial f_2}{\partial x_1} & \frac{\partial f_2}{\partial x_2} & \cdots & \frac{\partial f_2}{\partial x_n} \\ \vdots & \vdots & \vdots & \vdots \\ \frac{\partial f_n}{\partial x_1} & \frac{\partial f_n}{\partial x_2} & \cdots & \frac{\partial f_n}{\partial x_n} \end{bmatrix}}_{\substack{\text{Jacobian evaluated at last approx.} \\ \text{for } x_1, x_2, \ldots, x_n}} \underbrace{\begin{bmatrix} \Delta x_1 \\ \Delta x_2 \\ \vdots \\ \vdots \\ \Delta x_n \end{bmatrix}}_{\substack{\text{Update to apply} \\ \text{to last approx.}}} = -\underbrace{\begin{bmatrix} f_1 \\ f_2 \\ \vdots \\ \vdots \\ f_n \end{bmatrix}}_{\substack{\text{Negative of residual} \\ \text{when last approx.} \\ \text{substituted into (3.15)}}}. \tag{3.16}$$

If a solution is simple (multiplicity one),[19] Newton's method will be quadratically convergent, that is, with a sufficiently good initial approximation, the number of correct digits will roughly double for each iteration.

Application of Newton's method to (3.13)**:** Second-order FD discretization of (3.13) gives, for $n = 1, 2, \ldots, N-1$,

$$F_n = \frac{1}{h^2}\left[y_{n-1} - 2y_n + y_{n+1}\right] - f\left(x_n, y_n, \frac{1}{2h}\left[-y_{n-1} + y_{n+1}\right]\right) = 0,$$

resulting in a Newton system of the structure

$$\left[\begin{array}{ccccccc} 1 & & & & & & \\ \hline \times & \times & \times & & & & \\ & \times & \times & \times & & & \\ & & \ddots & \ddots & \ddots & & \\ & & & \times & \times & \times & \\ & & & & \times & \times & \times \\ \hline & & & & & & 1 \end{array}\right] \begin{bmatrix} \Delta y_0 \\ \Delta y_1 \\ \Delta y_2 \\ \vdots \\ \vdots \\ \Delta y_{N-1} \\ \Delta y_N \end{bmatrix} = -\left[\begin{array}{c} 0 \\ \hline F_1 \\ F_2 \\ \vdots \\ \vdots \\ F_N \\ \hline 0 \end{array}\right] \begin{array}{c} \text{Assume } y_0 \text{ correct} \\ \\ \text{Negative of} \\ \text{residual at} \\ \text{interior points} \\ \\ \text{Assume } y_N \text{ correct} \end{array}, \tag{3.17}$$

In the coefficient matrix, line n is $\left[\frac{\partial F_n}{\partial y_{n-1}}, \frac{\partial F_n}{\partial y_n}, \frac{\partial F_n}{\partial y_{n+1}}\right]$. The structure of (3.17) is very similar to (3.14), with one difference being that the matrix entries change slightly as the Newton iterations progress. If this Newton approach is applied to a linear ODE, it becomes equivalent to (3.14) and converges in one iteration, irrespective of the initial approximation.

3.3.2 Approaches for Increased Order of Accuracy

All three approaches below apply to both ODEs and PDEs.

[19] and the functions are twice continuously differentiable.

Richardson extrapolation: With all approximations centered, as just described, error expansions will contain even powers only in h, and Richardson extrapolation (Section E.1) will be particularly effective (gaining for each step two orders of accuracy). Using higher-order accurate derivative approximations can become problematic for a couple of reasons: (i) FD approximations of both the ODE and the BC may require additional ghost points, leading to awkward algebra; and (ii) the resulting systems of equations may be singular (as $h \to 0$) even for ODEs which have unique solutions.

Deferred correction: For a survey (historical and numerical) of this topic,[20] see Ong and Spiteri (2020).[21] Another pioneering paper is Pereyra (1966). The idea underlying this approach[22] is *iterative improvement*, as illustrated in the following example.

Example 3.3.1 Iteratively obtain an accurate solution to the linear system $A\underline{x} = \underline{b}$ if there is a fast algorithm available to solve the system approximately.

Solution Let $\underline{\tilde{x}}$ be an approximate solution. Then evaluate $A\underline{\tilde{x}} - \underline{b} = \underline{\eta}$ (where presumably $||\underline{\eta}|| << ||\underline{b}||$). Subtracting the exact $A\underline{x} = \underline{b}$ (with $\underline{x}$ unknown) gives $A(\underline{\tilde{x}} - \underline{x}) = \underline{\eta}$. Solving this last system gives an approximation for $\underline{\tilde{x}} - \underline{x}$, that is, a correction to the previous approximation $\underline{\tilde{x}}$. If necessary, the iteration can be repeated. □

Even with a quite inaccurate (e.g., low-order accurate) linear system solver, this approach gives an accuracy order matching that of the residual evaluation.[23] To address the problems (i), (ii) mentioned earlier when motivating Richardson extrapolation, the deferred correction idea is to use a second-order matrix for the solution step together with higher-order accurate derivative approximations[24] when computing the residual.

Compact stencils: Examples 1.2.2 and 1.2.3 illustrated the difference between explicit and implicit (compact) approximations in the case of $f''(x)$. Two orders of accuracy were gained by including three rather than just one f'' entry.

Example 3.3.2 Discretize the ODE BVP $y''(x) = f(x)$, $y(a) = \alpha$, $y(b) = \beta$ using explicit and implicit 3-node wide FD stencils.

[20] Also known as defect correction.
[21] The topic originated in its modern form in 1947 Fox (1947).
[22] and also underlying multigrid methods for solving (primarily) equilibrium-type (elliptic) PDEs.
[23] Assuming the linear system is not severely ill conditioned.
[24] maybe with noncentered FD approximations near boundaries.

Solution Following the description in Section 3.3.1.1, the explicit discretization becomes

$$\frac{1}{h^2}\left[\begin{array}{cccccccc} 1 & & & & & & & \\ \hline 1 & -2 & 1 & & & & & \\ & 1 & -2 & 1 & & & & \\ & & \ddots & \ddots & \ddots & & & \\ & & & 1 & -2 & 1 & & \\ & & & & 1 & -2 & 1 & \\ \hline & & & & & & 1 & \end{array}\right]\left[\begin{array}{c} y_0 \\ y_1 \\ y_2 \\ \vdots \\ \vdots \\ y_{N-1} \\ y_N \end{array}\right]=\left[\begin{array}{c} \alpha \\ \hline f(x_1) \\ f(x_2) \\ \vdots \\ \vdots \\ f(x_{N-1}) \\ \hline \beta \end{array}\right] \begin{array}{l} \text{Left BC} \\ \\ \text{ODE at} \\ \text{interior} \\ \text{points} \\ \\ \text{Right BC} \end{array}, \tag{3.18}$$

with its implicit counterpart differing only by the central section of the RHS vector having been replaced by

$$\frac{1}{12}\left[\begin{array}{ccccccc} 1 & 10 & 1 & & & & \\ & 1 & 10 & 1 & & & \\ & & \ddots & \ddots & \ddots & & \\ & & & 1 & 10 & 1 & \\ & & & & 1 & 10 & 1 \end{array}\right]\left[\begin{array}{c} f(x_0) \\ f(x_1) \\ f(x_2) \\ \vdots \\ \vdots \\ f(x_{N-1}) \\ f(x_N) \end{array}\right]. \tag{3.19}$$

Since the $f(x_i)$-values are assumed to be known, the cost increase is minimal. Two orders of accuracy have been gained in the solution of the ODE problem – from $O(h^2)$ to $O(h^4)$. □

A shortcoming of the compact stencil approach is apparent already in 1-D: The stencil weights do not depend linearly on the derivatives that are approximated. For example, to approximate $\alpha f'(x) + \beta f''(x)$ by explicit stencils, one simply finds weights for $f'(x)$ and $f''(x)$ and adds these with multipliers α and β, respectively. This linearity is lost for compact stencils. Even in the case of one term only but with a variable coefficient, say $\alpha(x) f'(x)$, compact stencils can become complicated.

Example 3.3.3 Illustrate the convergence rates when approximating the ODE

$$y''(x) = f(x) \quad \text{with} \quad f(x) = \cos(20\sqrt{x}), \quad \text{and} \quad y(0) = y(1) = 0 \tag{3.20}$$

to fourth-order accuracy with use of Richardson extrapolation, deferred correction, and compact stencils, respectively (in all cases solving only tri-diagonal linear systems).[25]

[25] The RHS function $f(x) = \cos(20\sqrt{x})$ will be used again later in the contexts of quadrature (Sections 8.3.3, F.2 and G.2.2) and is displayed in Figure 8.3(a).

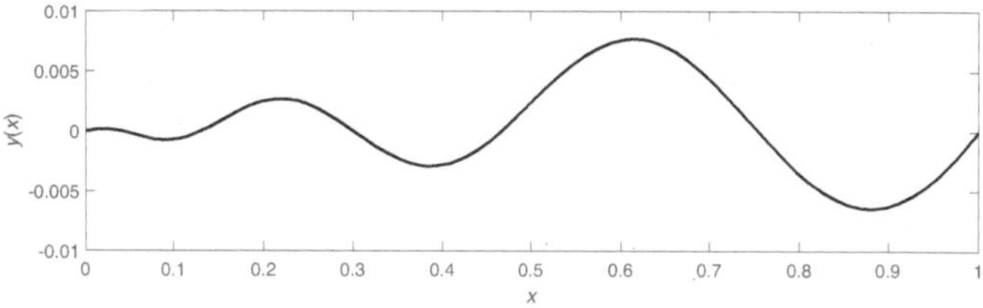

Figure 3.6 Analytic solution to (3.20):

$$y(x) = ((3 - 400x)\cos(20\sqrt{x}) + x(3 + 397\cos 20 - 60\sin 20) + 60\sqrt{x}\sin(20\sqrt{x}))/40000.$$

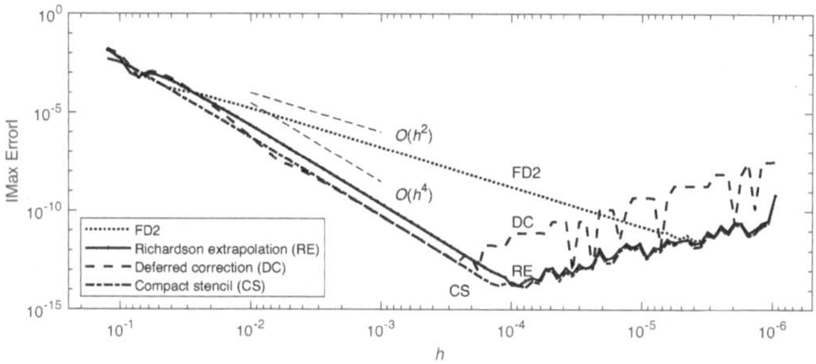

Figure 3.7 Convergence rates for the problem in Example 3.3.3. The three fourth-order approaches are contrasted against basic second-order FD2 approximation.

Solution The analytic solution to (3.20) is shown in Figure 3.6. Figure 3.7 shows for all the methods the expected convergence rates up to the point where double precision rounding errors enter. The first two approaches for increased accuracy (Richardson extrapolation and deferred correction) can readily be brought to still higher orders by simply being applied repeatedly. Some implementation comments:

Richardson extrapolation: The results for a given value of N combine results of using N and $N/2$ (i.e., step sizes $h = 1/N$ and $h = 2/N$) in the style of ((E.3).

Deferred correction: The residual vector (RHS) in the second solution stage was calculated with the fourth-order accurate approximation

$$\begin{bmatrix} r_1 \\ r_2 \\ r_3 \\ \\ \\ r_{N-2} \\ r_{N-1} \end{bmatrix} =$$

$$\frac{1}{h^2}\begin{bmatrix} \frac{5}{6} & -\frac{5}{4} & -\frac{1}{3} & \frac{7}{6} & -\frac{1}{2} & \frac{1}{12} & & & & & \\ -\frac{1}{12} & \frac{4}{3} & -\frac{5}{2} & \frac{4}{3} & -\frac{1}{12} & & & & & & \\ & -\frac{1}{12} & \frac{4}{3} & -\frac{5}{2} & \frac{4}{3} & -\frac{1}{12} & & & & & \\ & & \ddots & \ddots & \ddots & \ddots & \ddots & & & & \\ & & & \ddots & \ddots & \ddots & \ddots & \ddots & & & \\ & & & & & -\frac{1}{12} & \frac{4}{3} & -\frac{5}{2} & \frac{4}{3} & -\frac{1}{12} & \\ & & & & & \frac{1}{12} & -\frac{1}{2} & \frac{7}{6} & -\frac{1}{3} & -\frac{5}{4} & \frac{5}{6} \end{bmatrix}\begin{bmatrix} y(x_0) \\ y(x_1) \\ y(x_2) \\ y(x_3) \\ \\ \\ y(x_{N-1}) \\ y(x_N) \end{bmatrix}$$

$$-\begin{bmatrix} f(x_1) \\ f(x_2) \\ f(x_3) \\ \\ \\ \\ f(x_{N-1}) \end{bmatrix}.$$

The top and bottom rows of this $(N-1) \times (N+1)$ matrix use off-center approximations (rather than centered ones) in order not to extend outside the interval $[x_0, x_N] = [0, 1]$.[26]

Compact stencil: This implementation strictly followed (3.18), (3.19). □

[26] The quite large rounding errors can be reduced by computing the residual vector (often a relatively inexpensive step) in extended precision arithmetic.

4
Grid-based FD Approximations for Partial Differential Equations

Finite difference approximations for partial differential equations (PDEs) are a vast topic that cannot be adequately covered in a single chapter. Many numerical analysis textbooks treat this subject extensively. Books specializing in FD approaches for ODEs and PDEs include Iserles (2012); Kreiss and Ortiz (2014); LeVeque (2007); Morton and Mayers (1994), while Strikwerda (2004) focuses only on PDEs.[1] This present chapter is best seen as a rough road map to the subject area, together with isolated observations regarding a few typically less well-covered issues (in particular concerning high-accuracy aspects). In almost all parts, we need to refer to more specialized literature.

Partial differential equations (or systems of equations including PDEs) that arise in applications often combine features that can be understood by considering much simpler "model" PDEs. Sections 4.1 and 4.2 of this chapter focus on simplified time-dependent PDEs, and Section 4.3 on time-independent PDEs. In both cases, boundaries play a significant role (especially if these are curved, and do not align with grid lines). Some FD-based approaches to deal with this issue are discussed in Section 4.4. A more general approach for handling both irregular boundaries and local refinement in select areas is provided by changing from grid-based to mesh-free FD discretizations, as described in Chapter 5.

Ill-posedness (solutions not depending continuously on the data provided) arises more often with PDEs than with ODEs and, especially for more complicated PDE systems, should be assessed before numerics is undertaken.

[1] In most cases on low-order accuracy approximations.

4.1 Time-Dependent PDEs

These are initial value problems (usually in bounded domains). Sections 4.1.1–4.1.3 focus on time stepping strategies, with Section 4.1.4 describing the idea of staggered grids.

4.1.1 Method of Lines (MOL)

The MOL concept is straightforward, and the approach is widely utilized (Schiesser, 1991). One first discretizes in all but one independent direction, typically in all space directions, leaving derivatives in time. Next, a standard ODE initial value solver is applied in this remaining direction. Advantages include great generality, simplicity in coding, and in readily reaching high orders of accuracy.[2] Stability can usually be assessed by comparing the eigenvalues of the spatial differentiation matrix with the ODE solver's stability domain (Reddy and Trefethen, 1992; Samarin and Chudov, 1963).

4.1.2 Combined Space-Time Discretizations

While MOL is the dominant time stepping approach, there are alternatives that can be advantageous in terms of better utilizing special features of governing equations.

The Lax–Wendroff method The 1-D wave equation

$$\frac{\partial u}{\partial t} = \frac{\partial u}{\partial x} \tag{4.1}$$

suffices to illustrate the Lax–Wendroff concept. The small 4-node stencil

$$\frac{u(x,t+k) - u(x,t)}{k} = \frac{u(x+h,t) - u(x-h,t)}{2h} \tag{4.2}$$

is an MOL example, using FE in time. It has two shortcomings: (i) it is only first-order accurate in time; and (ii) as we will see in Example 4.1.2, time stepping with it becomes unconditionally unstable.

Taylor expansion of the error in (4.2) gives

$$\frac{u(x,t+k) - u(x,t)}{k} - \frac{u(x+h,t) - u(x-h,t)}{2h} = \frac{k}{2}u_{tt} + O(k^2) + O(h^2). \tag{4.3}$$

[2] In RK methods, the internal stages are only first- or second-order accurate, and the resulting high order of accuracy is only reached by precise cancellations of such errors when the stages are combined to form a complete time step. Using exact spatial boundary conditions during the internal stages can cause incompatibilities with these expected errors, requiring some special attention; see Carpenter et al. (1995); Pathria (1997); or Fornberg (1996), Section G1c.

From (4.1) it follows that $u_{tt} = (u_x)_t = (u_t)_x = u_{xx}$. The first term in the RHS of (4.3) can thus be approximated by $\frac{k}{2h^2}(u(x-h,t) - 2u(x,t) + 2u(x+h,t))$, resulting in an approximation that again fits into the same 4-node stencil shape

$$\frac{u(x,t+k) - u(x,t)}{k} = \frac{u(x+h,t) - u(x-h,t)}{2h} + \frac{u(x-h,t) - 2u(x,t) + u(x+h,t)}{kh^2}, \tag{4.4}$$

but is second-order accurate in both x and t. Furthermore, as seen in Example 4.1.3, it is stable under the largest time steps that the CFL condition[3] allows.

This Lax–Wendroff approach generalizes to higher accuracy orders (Gustafsson, 2008) and applies also to compressible flow problems (Morton and Mayers, 1994). Additionally, it can permit grid staggering, as described in Section 4.1.4.

4.1.3 Stability Analysis

For ODEs, zero stability together with consistency (the latter trivially true for any method first-order accurate or better) sufficed for assuring convergence as the time step $k \to 0$. For time stepping PDEs, the corresponding *Lax Equivalence theorem*[4] also requires linearity:

Theorem 4.1.1 *For a well-posed linear problem, a consistent linear approximation is convergent if and only if it is stable.*

This again makes stability the crucial quantity to test for. Boundary conditions, variable coefficients, and nonlinearities all complicate testing for stability. We next focus on the highly simplified case of constant coefficient pure initial-value problems. Stability in these cases gives insights into what might be expected in other cases. Without attempting any rigor, we illustrate the methods by examples.

4.1.3.1 Von Neumann Analysis

The key idea is that any initial condition on an h-spaced grid without boundaries can be seen as a superposition of Fourier modes $e^{i\omega x}$ with $-\frac{\pi}{h} \le \omega \le \frac{\pi}{h}$. For stability, one needs to ensure that none of these modes can grow arbitrarily fast as time is advanced. This often leads to a condition on the ratio of the time step k and a power of the space step h.

[3] Section 4.1.3.3
[4] Published in collaboration with R.D. Richtmyer (Lax and Richtmyer, 1956).

Example 4.1.1 Consider the FD scheme

$$\frac{u(x,t+k)-u(x,t)}{k}=\frac{u(x-h,t)-2u(x,t)+u(x+h,t)}{h^2} \tag{4.5}$$

for the heat equation $u_t = u_{xx}$. What is the stability condition for this scheme?

Solution Applied to a general Fourier mode $e^{i\omega x}$ in space, this mode can only change by some factor σ for each time step. Thus, one routinely substitutes $u(x,t) = \sigma^{t/k}e^{i\omega x}$ into the difference scheme. After that, $\sigma^{t/k}e^{i\omega x}$ can be factored out again, giving (in the case of (4.5))

$$\frac{\sigma-1}{k}=\frac{e^{-i\omega h}-2+e^{i\omega h}}{h^2}$$

and

$$\sigma = 1-4\frac{k}{h^2}\left(\sin\frac{\omega h}{2}\right)^2. \tag{4.6}$$

The factor $\left(\sin\frac{\omega h}{2}\right)^2$ can take any value in the range $[0,1]$. As σ is the growth factor for each time step and, under refinement, the number of steps needed increases indefinitely, stability requires that $|\sigma| \le 1$. The stability condition $k/h^2 \le \frac{1}{2}$ follows then from (4.6). □

Figures 4.1(a), (b) show the result of time stepping the scheme (4.5) with $h = 0.05$ together with $k = 0.45h^2$ and $k = 0.55h^2$, respectively. In both cases, the initial condition is $u(x,0) = 1$ and the boundary conditions are $u(0,t) = u(1,t) = 0$ for $t > 0$. The first case satisfies the stability condition $k/h^2 \le \frac{1}{2}$, while the second case violates it.

Example 4.1.2 Determine the stability condition of the FD scheme (4.2).

Solution The same procedure as in the previous example now gives $\frac{\sigma-1}{k} = \frac{e^{i\omega h}-e^{-i\omega h}}{2h}$ and $\sigma = 1 + i\frac{k}{h}\sin\omega h$. No combination of $k > 0$ and $h > 0$ will ensure $|\sigma| \le 1$. The scheme is unconditionally unstable. □

Example 4.1.3 Determine the stability of the Lax–Wendroff scheme (4.4).

Solution The procedure above, with $\lambda = \frac{k}{h}$, gives, after some simplifications, $\sigma = 1+i\lambda\sin\omega h+\lambda^2(\cos\omega h-1)$ and $|\sigma|^2 = 1-\lambda^2(1-\lambda^2)(1-\cos\omega h)^2$. With $(1-\cos\omega h)^2$ varying over the range $[0,4]$, stability requires $0 \le \lambda^2(1-\lambda^2) \le \frac{1}{2}$, which is satisfied for $\lambda = \frac{k}{h} \le 1$. □

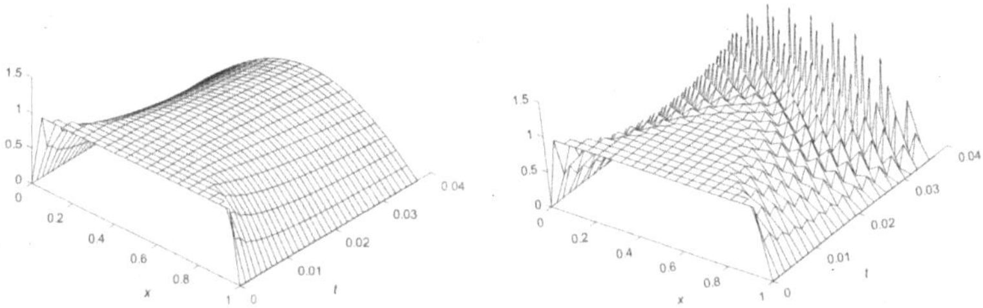

(a) $k/h^2 = 0.45 < \frac{1}{2}$. (b) $k/h^2 = 0.55 > \frac{1}{2}$.

Figure 4.1 Numerical solutions when time stepping the heat equation $u_t = u_{xx}$ with the scheme (4.5) using time steps just below and just above the stability condition $k/h^2 \le \frac{1}{2}$.

Example 4.1.4 Determine the stability of the leapfrog-FD2 scheme:

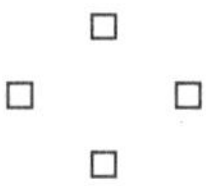

$$\frac{u(x,t+k) - u(x,t-k)}{2k} = \frac{u(x+h,t) - u(x-h,t)}{2h}. \quad (4.7)$$

Solution The routine substitutions $u(x,t) = \sigma^{t/k} e^{i\omega x}$ and $\lambda = \frac{k}{h}$ give

$$\sigma^2 + (2i \underbrace{\lambda \sin \omega h}_{\alpha})\sigma - 1 = 0,$$

with roots $\sigma_{1,2} = -i\alpha \pm \sqrt{1-\alpha^2}$. There are now three cases:

Case 1. $\lambda < 1$: In this case, $|\alpha| = |\lambda \sin \omega h| < 1$, and the two roots σ_1, σ_2 are separate, both with magnitude one. The solution $u(x,t)$ evolves as $u(x,t) = \left(A\,\sigma_1^{t/k} + B\,\sigma_2^{t/k}\right) e^{i\omega x}$, which cannot grow in time. The scheme is stable.

Case 2. $\lambda = 1$: There is now a possibility of σ being a double root on the unit circle, arising when $\sin \omega h = \pm 1$ and $\omega = \pm\frac{\pi}{2h}$.[5] In this case $e^{i\omega x}$ (real or imaginary parts) oscillate with the pattern

$$\cdots \quad 0 \quad -1 \quad 0 \quad 1 \quad 0 \quad -1 \quad 0 \quad 1 \quad 0 \quad -1 \quad 0 \quad 1 \quad \cdots$$

[5] The assumed solution form $u(x,t) = \sigma^{t/k} e^{i\omega x}$ can only directly represent cases with exponential rather than linear changes in time, as arises for roots σ of magnitude 1 and multiplicity above 1.

revealing for grid points in the x, t-plane the pattern (when initializing with all zeros on the time level below)

$$
\begin{array}{cccccccccccccc}
 & \vdots & \vdots & \vdots & \vdots & \vdots & \vdots & \vdots & \vdots & \vdots & \vdots & \vdots & \vdots & \\
\cdots & 0 & -5 & 0 & 5 & 0 & -5 & 0 & 5 & 0 & -5 & 0 & 5 & \cdots \\
\cdots & 4 & 0 & -4 & 0 & 4 & 0 & -4 & 0 & 4 & 0 & -4 & 0 & \cdots \\
\cdots & 0 & 3 & 0 & -3 & 0 & 3 & 0 & -3 & 0 & 3 & 0 & -3 & \cdots \\
\cdots & -2 & 0 & 2 & 0 & -2 & 0 & 2 & 0 & -2 & 0 & 2 & 0 & \cdots \\
\cdots & 0 & -1 & 0 & 1 & 0 & -1 & 0 & 1 & 0 & -1 & 0 & 1 & \cdots \\
\cdots & 0 & 0 & 0 & 0 & 0 & 0 & 0 & 0 & 0 & 0 & 0 & 0 & \cdots
\end{array}
$$

This grows without bound under refinement, making the scheme unstable.

Case 3. $\lambda > 1$: There will be a possibility of $\alpha > 1$ (or $\alpha < -1$), for example, again for $\omega = \pm\frac{\pi}{2h}$. One of the roots $\sigma_{1,2} = -i\alpha \pm \sqrt{1-\alpha^2}$ will be outside the unit circle, implying divergence. □

As a follow-up to Example 4.1.4, we recognize in the $\alpha = \lambda \sin \omega h$ substitution the same factor $\sin \omega h$ as seen before in (2.3) and Figure 2.1. If we change FD2 in space to FD$\{p\}$ (with p even), keeping leapfrog in time, the stability condition similarly becomes

$$\lambda < \frac{1}{\text{max value of the corresponding curve in Figure 2.1}}.$$

In particular, increasing the spatial accuracy indefinitely just reduces the condition on $\lambda = \frac{k}{h}$ by a factor of π. This last observation carries over to all MOL time stepping schemes (such as LM, RK, etc.).

4.1.3.2 Stability Domain-based Analysis for MOL Methods

This analysis approach is very flexible and can be used both analytically (as here for grid-based FD methods) and numerically (for example, in RBF contexts, see Chapter 5, with the eigenvalues of the differentiation matrix (DM) computed numerically). The idea is that all eigenvalues of the DM must fall within the ODE solver's stability domain. In ODE contexts (when the number of equations of an ODE system is fixed), the solver's stability domain shows if there are growing solutions, or not. In PDE MOL contexts, the number of ODEs increases under refinement, and the stability domain instead settles convergence versus divergence.[6] The three examples below illustrate the approach for grid-based FD MOL methods.

[6] Corresponding to what the zero-stability criterion does for ODE solvers.

Example 4.1.5 Determine the stability condition for the FE-FD2 scheme in Example 4.1.1.

Solution In a periodic setting, the ODE system becomes

$$\frac{d}{dt}\begin{bmatrix} u_1(t) \\ u_2(t) \\ \vdots \\ \vdots \\ u_n(t) \end{bmatrix} = \frac{1}{h^2}\begin{bmatrix} -2 & 1 & & & 1 \\ 1 & -2 & 1 & & \\ & \ddots & \ddots & \ddots & \\ & & \ddots & \ddots & \ddots \\ 1 & & & 1 & -2 \end{bmatrix}\begin{bmatrix} u_1(t) \\ u_2(t) \\ \vdots \\ \vdots \\ u_n(t) \end{bmatrix}.$$

The eigenvalues λ of the DM are real, and fall in the interval $[-4, 0]/h^2$.[7] This interval needs to fall within the FE stability domain, which for $\xi = k\lambda$ (real) is $[-2, 0]$ (cf. the AB1 = FE case in Figure 3.5(a)). It thus follows that $k/h^2 \leq \frac{1}{2}$ (as obtained before when using von Neumann analysis in Example 4.1.1). □

Example 4.1.6 Determine the stability condition for the FE-FD2 scheme in Example 4.1.2.

Solution The DM is in this case

$$\frac{1}{2h}\begin{bmatrix} 0 & 1 & & & -1 \\ -1 & 0 & 1 & & \\ & \ddots & \ddots & \ddots & \\ & & \ddots & \ddots & \ddots \\ 1 & & & -1 & 0 \end{bmatrix},$$

with eigenvalues purely imaginary, along the interval $\frac{i}{h}[-1, 1]$. The FE stability domain has no imaginary axis coverage, implying this scheme is unconditionally unstable. □

Example 4.1.7 Determine the stability condition for the leapfrog-FD2 scheme in Example 4.1.4.

Solution The DM is the same as in Example 4.1.6, and one readily shows that the leapfrog stability domain is exactly the interval $i\,[-1, 1]$ along the imaginary ξ-axis. The stability condition thus becomes $k/h \leq 1$.[8] □

In the three examples, Gershgorin's theorem (see standard linear algebra texts) gave the precise eigenvalue bounds (rather than unnecessarily large

[7] As the matrix is symmetric, with eigenvalues bound by Gershgorin's theorem.

[8] This analysis does not catch the subtlety that we ought to exclude the case of k/h exactly equal to 1.

ones, leading to more restrictive stability conditions than necessary). However, in the FD4 case, the DM (cf. the second line in Table 1.1) becomes

$$\frac{1}{h}\begin{bmatrix} 0 & \frac{2}{3} & -\frac{1}{12} & & & \frac{1}{12} & -\frac{2}{3} \\ -\frac{2}{3} & 0 & \frac{2}{3} & -\frac{1}{12} & & & \frac{1}{12} \\ \frac{1}{12} & -\frac{2}{3} & 0 & \frac{2}{3} & -\frac{1}{12} & & \\ & \ddots & \ddots & \ddots & \ddots & \ddots & \\ & & \ddots & \ddots & \ddots & \ddots & \ddots \\ -\frac{1}{12} & & & \frac{1}{12} & -\frac{2}{3} & 0 & \frac{2}{3} \\ \frac{2}{3} & -\frac{1}{12} & & & \frac{1}{12} & -\frac{2}{3} & 0 \end{bmatrix}.$$

Gershgorin's theorem gives here the bound $\frac{i}{h}[-\frac{3}{2},+\frac{3}{2}]$, versus the sharp bound of approximately $\frac{i}{h}[-1.372,+1.372]$.[9]

Example 4.1.8 Determine the stability condition for a RK4–FD4 approximation of the heat equation $u_t = u_{xx}$.

Solution We obtain from the second line in Table 1.2 that

$$u_{xx} \approx \left[\ -\tfrac{1}{2} \quad \tfrac{4}{3} \quad -\tfrac{5}{2} \quad \tfrac{4}{3} \quad -\tfrac{1}{2}\ \right] u/h^2.$$

When applied to $e^{i\omega x}$, this gives $\frac{2}{3}(\cos(\omega h) - 7)\left(\sin\frac{\omega h}{2}\right)^2/h^2$, which takes values within $\left[-\frac{16}{3},0\right]/h^2$. This interval has to fit into the RK4 method's real axis coverage, which is approximately $[-2.78529, 0]/k$, that is, the stability condition becomes $k/h^2 \leq 0.52224$.[10] □

4.1.3.3 Courant–Friedrichs–Lewy (CFL) Condition

For a wave (advection) equation approximated by an explicit FD stencil, there is a geometric upper limit on how fast information can travel. For example, with a stencil shaped in the x,t-plane as $\begin{array}{ccc} & \square & \\ \square & \square & \square \end{array}$, information can travel no more than h sideways when time is advanced by k, giving a maximum numerical velocity of h/k. The CFL condition asserts that this must not be less than an actual characteristic velocity for the PDE. For the PDE $u_t = \alpha\, u_x$, the condition

[9] The constant can be read off from the peak value $\sqrt{\frac{1}{4}+2\sqrt{\frac{2}{3}}} \approx 1.372$ of the FD4 curve in Figure 2.1.

[10] Computations are normally most cost efficient when the errors from different sources roughly match. This is not the case in this example (intended only to illustrate stability analysis). With $k/h^2 <$ constant, second-order accuracy would have sufficed in time.

thus becomes $k/h \leq 1/|\alpha|$. This CFL test will never ensure stability, but it ensures instability if it is violated.

The CFL condition was described already in 1928 (Courant et al., 1928), yet the concept of numerical stability for FD schemes was not well understood before the work by von Neumann a couple of decades later (Lax, 1967).[11]

4.1.3.4 Boundary Effects, Variable Coefficients, and Nonlinear Instabilities

In all three cases listed in this section header, the simplified assumptions required for von Neumann analysis are violated. Strict stability analysis becomes algebraically more difficult (if at all possible), and we need to refer to more specialized literature, for example, Gustafsson (2008); Gustafsson et al. (2013).

Energy methods can sometimes provide strict results. From a practical perspective, locally changing to constant coefficients and analyzing these cases give good guidelines for stability also in more complicated cases. Addition of small amounts of artificial viscosity (or hyperviscosity, as is commonly used in RBF–FD contexts, see Section 5.4.3) adds extra safety. Figures 4.5-2, 4.5-3, and 4.6-4 in Fornberg (1996) illustrate typical divergence patterns for (i) violation of von Neumann condition, (ii) boundary-induced instability, and (iii) nonlinear instability. As seen in Figure 4.1, instabilities for case (i) typically take the form of growing high-frequency oscillations across some interval, whereas instabilities for case (ii) may be most notable at or near to boundaries. Nonlinear instabilities (case (iii)) take often the form of localized spikes or oscillations. The method of *Summation By Parts* (SBP), in particular in its Simultaneous Approximation Term (SAT) form, provides for many PDEs a systematic way to obtain guaranteed stable boundary conditions (Gustafsson, 2008; Svärd and Nordström, 2014). The approach can be applied in multiple dimensions and for general function spaces (Glaubitz et al., 2023).

4.1.4 Staggered Grids

4.1.4.1 Concept

Table 4.1 should be compared to Table 1.1. The approximations are accurate in both cases at the stencil center, but here this center falls half-way between grid points rather than at a grid point. The bottom row labeled "limit" shows a more rapid sideways decrease in magnitude, as $O(1/n^2)$ rather than $O(1/n)$, where n denotes the distance to the stencil center point. The leading error term in approximations with the same accuracy order is also smaller, as seen

[11] This time delay was partly caused by the original description being in the different context of existence proofs of solutions.

Table 4.1 *Weights for centered staggered FD approximations of the first derivative on an equispaced grid, accurate at the stencils' midpoints (half-way between grid points), (omitting the factor* $1/h$*).*

Order					Weights First derivative						
2					-1	1					
4				$\frac{1}{24}$	$-\frac{9}{8}$	$\frac{9}{8}$	$-\frac{1}{24}$				
6			$-\frac{3}{640}$	$\frac{25}{384}$	$-\frac{75}{64}$	$\frac{75}{64}$	$-\frac{25}{384}$	$\frac{3}{640}$			
8		$\frac{5}{7168}$	$-\frac{49}{5120}$	$\frac{245}{3072}$	$-\frac{1225}{1024}$	$\frac{1225}{1024}$	$-\frac{245}{3072}$	$\frac{49}{5120}$	$-\frac{5}{7168}$		
10	$-\frac{35}{294912}$	$\frac{405}{229376}$	$-\frac{567}{40960}$	$\frac{735}{8192}$	$-\frac{19845}{16384}$	$\frac{19845}{16384}$	$-\frac{735}{8192}$	$\frac{567}{40960}$	$-\frac{405}{229376}$	$\frac{35}{294912}$	
$\vdots$	$\downarrow$	$\downarrow$	$\downarrow$	$\downarrow$	$\downarrow$	$\downarrow$	$\downarrow$	$\downarrow$	$\downarrow$	$\downarrow$	
Limit $\cdots$	$-\frac{4}{9^2\pi}$	$\frac{4}{7^2\pi}$	$-\frac{4}{5^2\pi}$	$\frac{4}{3^2\pi}$	$-\frac{4}{1^2\pi}$	$\frac{4}{1^2\pi}$	$-\frac{4}{3^2\pi}$	$\frac{4}{5^2\pi}$	$-\frac{4}{7^2\pi}$	$\frac{4}{9^2\pi}$	$\cdots$

in Table 4.2. This is not surprising, since interpolants in general feature their largest errors in slope at grid points (see, for example, Figures 1.3, 1.4, and 2.3).

Early applications of grid staggering include Harlow and Welch (1965) (while introducing the "marker and cell" method for 3-D flows) and Yee (1966) (the Yee scheme) for the 3-D Maxwell's equations (see Section 4.1.4.2). These schemes correspond to utilizing the top row (order $p = 2$) from Table 4.1. The concept of staggering generalizes to derivatives of all orders,[12] larger classes of PDEs, and to mesh-free situations (cf. Section 5.4.5). It is also a key ingredient in finite volume methods (LeVeque, 2002).

The benefits of interlacing in space the locations where $f(x)$ is given and $f'(x)$ is sought (or where different dependent functions are represented) also carry over to the time direction. As in space, the benefits of time staggering increase with increasing orders of accuracy. Time-staggered generalizations of linear multistep and Runge–Kutta methods are discussed in Ghrist et al. (2000); Verwer (2007).

[12] Most advantageously for odd orders; for even orders, nonstaggered approximations are typically more accurate.

Table 4.2 *The leading error terms in the regular and staggered first derivative approximations in Tables 1.1 and 4.1, respectively. The ratio* $0.2500 = 1/4$ *in the case* $p = 2$ *can be viewed as using the same second-order accurate approximation with h halved. The accuracy advantage with staggering is seen to then increase with the stencil size/accuracy order.*

Accuracy of order p	2	4	6	8	...
Type of leading error term	$h^2 f^{(3)}(x)$	$h^4 f^{(5)}(x)$	$h^6 f^{(7)}(x)$	$h^8 f^{(9)}(x)$	...
Coefficient for leading error term					
Regular grid	$\frac{1}{6}$	$-\frac{1}{30}$	$\frac{1}{140}$	$-\frac{1}{630}$	...
Staggered grid	$\frac{1}{24}$	$-\frac{3}{640}$	$\frac{5}{7168}$	$-\frac{35}{294912}$	...
Approximate ratio	0.2500	0.1406	0.0977	0.0748	...

4.1.4.2 Illustrations of Grid Staggering for Three Types of Wave Equations

We use three systems of wave equations to illustrate the concept of grid staggering:

1-D acoustic

$$\begin{cases} \frac{\partial u}{\partial t} = c\frac{\partial v}{\partial x} \\ \frac{\partial v}{\partial t} = c\frac{\partial u}{\partial x} \end{cases}$$

2-D elastic

$$\begin{cases} \rho\frac{\partial u}{\partial t} = \frac{\partial f}{\partial x} + \frac{\partial g}{\partial y} \\ \rho\frac{\partial v}{\partial t} = \frac{\partial g}{\partial x} + \frac{\partial h}{\partial y} \\ \frac{\partial f}{\partial t} = (\lambda + 2\mu)\frac{\partial u}{\partial x} + \lambda\frac{\partial v}{\partial y} \\ \frac{\partial g}{\partial t} = \mu\left(\frac{\partial v}{\partial x} + \frac{\partial u}{\partial y}\right) \\ \frac{\partial h}{\partial t} = \lambda\frac{\partial u}{\partial x} + (\lambda + 2\mu)\frac{\partial v}{\partial y} \end{cases}$$

3-D Maxwell

$$\begin{cases} \frac{\partial E_x}{\partial t} = \frac{1}{\varepsilon}\left(\frac{\partial H_z}{\partial y} - \frac{\partial H_y}{\partial z}\right) \\ \frac{\partial E_y}{\partial t} = \frac{1}{\varepsilon}\left(\frac{\partial H_x}{\partial z} - \frac{\partial H_z}{\partial x}\right) \\ \frac{\partial E_z}{\partial t} = \frac{1}{\varepsilon}\left(\frac{\partial H_y}{\partial x} - \frac{\partial H_x}{\partial y}\right) \\ \frac{\partial H_x}{\partial t} = \frac{1}{\mu}\left(\frac{\partial E_y}{\partial z} - \frac{\partial E_z}{\partial y}\right) \\ \frac{\partial H_y}{\partial t} = \frac{1}{\mu}\left(\frac{\partial E_z}{\partial x} - \frac{\partial E_x}{\partial z}\right) \\ \frac{\partial H_z}{\partial t} = \frac{1}{\mu}\left(\frac{\partial E_x}{\partial y} - \frac{\partial E_y}{\partial x}\right) \end{cases}$$

The systems are here all written in terms of first derivatives (rather than fewer equations using second derivatives).[13,14]

[13] First-order formulations are often closer to how governing equations are arrived at from basic physical principles.

[14] It depends on the details of a PDE system of equations (which first derivatives appear and in what combinations) if staggering is applicable. However, remarkably often, this turns out to be the case.

1-D acoustic wave equation: Figure 4.2(a) shows the most straightforward grid in x, t-space, with each of the two variables u, v represented at each grid point. However, if one uses centered second-order FD in both space and time, this calculation amounts to four entirely independent calculations. Figure 4.2(b) illustrates one of these four. Benefits include: (i) a quarter of the computational cost and storage; and (ii) if the four calculations drift somewhat apart, this would appear as high-frequency oscillations across the grid, which then need to be dealt with. Figure 4.3 matches Figure 4.2(b), but with additional markers (inside dashed boxes) indicating locations where time derivatives for the respective variables are centered.

2-D elastic wave equation: In this PDE system, u and v denote the horizontal and vertical velocities of material particles, and f, g, h (sometimes denoted $\Gamma_{1,1}, \Gamma_{1,2}, \Gamma_{2,2}$) are the three components of the (symmetric) stress tensor. The (spatially varying) material parameters ρ, λ, μ are typically assumed to be given. This system of equations incorporates a variety of wave phenomena, such as pressure (P) waves (with velocity $\sqrt{(\lambda + 2\mu)/\rho}$), shear ($S$) waves (with velocity $\sqrt{\mu/\rho}$), Rayleigh and Stoneley waves (following free surfaces and internal interfaces, respectively), waves striking interfaces at arbitrary angles giving rise to outgoing reflected and transmitted waves. In the same style of notation as used in Figure 4.3, Figure 4.4 shows how the five dependent variables in this case can be laid out to again allow staggering in both space and time.

3-D Maxwell's equations: Maxwell's equations form a foundation of modern physics (including for relativity theory). A variety of formulations are available, applicable to electromagnetic waves in the presence of currents and charges, traveling through spatially variable media. For our present purpose of illustrating 3-D grid staggering, we consider for simplicity the system of six PDEs shown earlier, describing how the three spatial components of an electric field E and a magnetic field H evolve in vacuum.

Grid staggering is again immediately applicable, with period $2h$ in all four dimensions (three in space plus time). Figure 4.5 shows over a $3 \times 3 \times 3$ node set how the six variables $E_x, E_y, E_z, H_x, H_y, H_z$ can be spatially arranged so that each variable can be updated by centered FD approximations. As the period box is $2 \times 2 \times 2$ (rather than $3 \times 3 \times 3$ used here for better graphical clarity), we see, for example, from the bottom $2 \times 2 \times 2$ block that in its eight positions, each variable appears only once. For staggering in time, we note further that the E- and H-entries can be present on alternating time levels, leading to a total node (memory) reduction by a factor of 16. For more information, especially on the

(a) Regular grid in space/time (b) Staggered grid in space/time

Figure 4.2 Layout of variables in the x, t-plane in the 1-D acoustic case. Part (b) shows one of the four uncoupled staggered schemes contained in the regular node layout in Part (a).

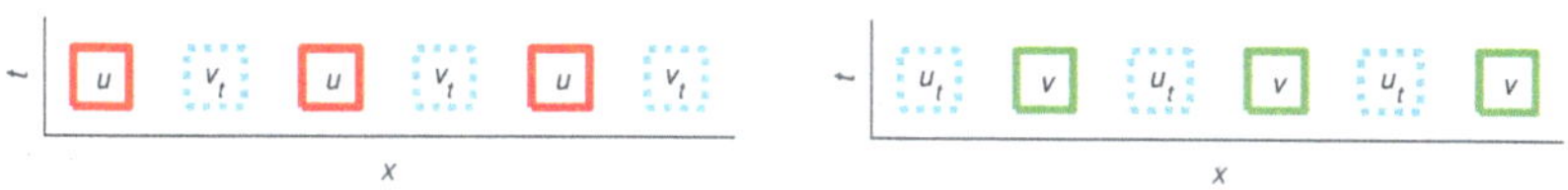

(a) Even time levels (b) Odd time levels

Figure 4.3 Same node layout for the 1-D acoustic equation as shown in Figure 4.2(b), but displayed more compactly.

pioneering Yee scheme (staggered second-order FD approximations in all four directions), see Kunz and Lubbers (1993); Taflove et al. (2005); Yee (1966).

4.2 Some Considerations When Approximating Convection-Type PDEs

Sections 4.2.1–4.2.4 give some schematic FD considerations for different categories of PDEs in which convection plays a major role. In all cases, some key observations can be made already with highly simplified model equations. Section 4.2.1 also illustrates the importance of using high orders of accuracy.

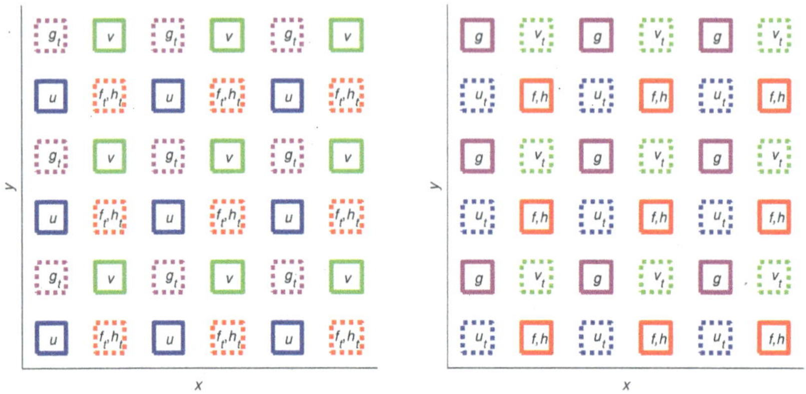

(a) Even time levels (b) Odd time levels

Figure 4.4 Layout of variables in the x, y-plane (solid line squares) and center locations for stencils for updating respective variables (dashed squares) in the 2-D elastic case.

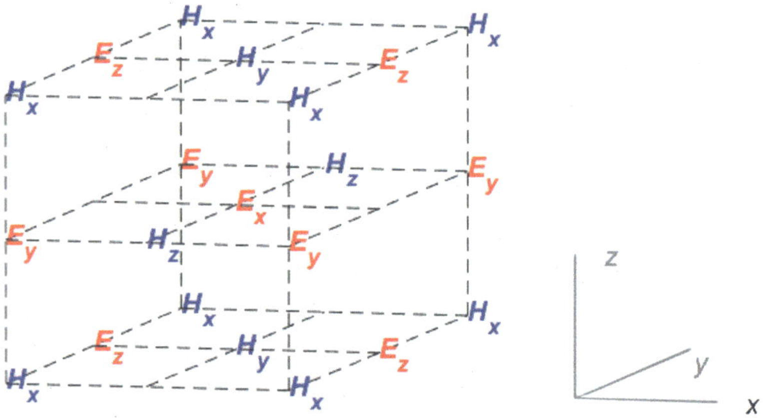

Figure 4.5 A $3 \times 3 \times 3$ block illustrating the staggered node layout for the 3-D Maxwell's equations ($2 \times 2 \times 2$ periodic; also with staggering available in time by alternating between the E- and the H-variables).

4.2.1 Linear Convection

The arguably simplest PDE for convection (advection, transport) is

$$\frac{\partial u}{\partial t} + \frac{\partial u}{\partial x} = 0, \tag{4.8}$$

together with suitable initial and boundary conditions.[15]

Figure 2.1 illustrates that, when using standard centered FD approximations in space, Fourier modes will generally travel too slowly (as compared to the straight line labeled "Exact (PS)," corresponding to all modes for (4.8) translating with their analytically correct uniform velocity). Higher modes traveling too slowly cause a pulse-type initial condition (as seen in Figure 4.6(a)) to develop trailing wave trains. These are in part (b) of the figure compared for FD2 and FD6 spatial approximations (lines labeled order 2 and order 6 in Table 1.1) at the time corresponding to when the pulse ideally should have translated 1,000 times the length h of the space step.[16]

This example dispels a commonly held misconception that high accuracy methods are unnecessary in application contexts where available data is of low accuracy (say, accurate to only one or two significant digits). One such situation arises in seismic exploration. In field testing, pressure pulses are initiated above a region of interest, and waves returning to the surface are recorded (after transmission through various layers and also partial reflections at interfaces). The industry standard when modeling this was in the 1970s to use second-order FD approximations (in 2-D), increased to about 20th-order at present (in 3-D, now instead making the handling of interfaces a dominant error source). Unless high orders are used, spurious trailing wave trains will severely corrupt the results from the numerical modeling. In other applications of modeling wave propagation, the task is to compute how pulses or waves get modified (maybe only weakly) due to smooth variations in the medium they travel through. Trailing wave trains, as seen in the FD2 case, are again extremely destructive. Bringing up the order of accuracy is the only generally applicable way to improve the integrity of such calculations.

The importance of high accuracy cannot readily be seen from the leading error terms in FD approximations. In the present example of FD2 and FD6 approximations, these are $\frac{1}{6}h^2 f^{(3)}(x)$ and $\frac{1}{140}h^6 f^{(7)}(x)$, respectively. In the case of a traveling discontinuity, the solution does not admit even a first derivative (and much less derivatives of orders 3 and 7). Nevertheless, the analysis in Fornberg (1996), Section 4.2, shows that the slope at the traveling discontinuity, ideally remaining infinite for all times, will for a pth-order FD approximation in

[15] Sometimes informally referred to as the Kreiss equation, due to the numerous insights H.-O. Kreiss arrived at, often with this equation as a starting point.

[16] The numerical time stepping was here carried out with an MOL approach, sufficiently accurate that all errors that are seen come from the space discretization. Improving the accuracy reached by a low-order FD scheme by means of reducing the space step h becomes often prohibitively costly (both in computer time and memory usage), especially in higher dimensions and when combined with explicit time stepping.

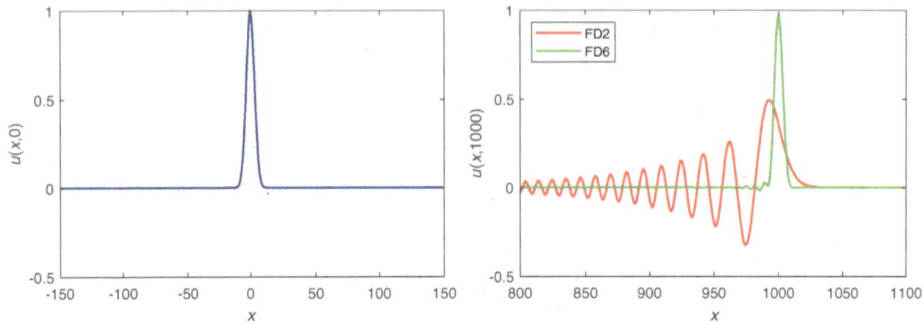

(a) Initial condition (b) Pulse after traveling 1,000 h

Figure 4.6 The FD2 vs. FD6 test case described in Section 4.2.1. The scale on the horizontal axes shows the number of grid spacings h.

space decay under error-free time integration as $O(\sqrt[p+1]{t})$, that is, there is also in this quite extreme case a vast advantage in using high-order FD approximations.

4.2.2 Convection–Diffusion Equations

Although the equation (4.8) models only the simplest form of pure convection in 1-D, it serves well for understanding many aspects of numerical schemes for convection-dominated PDEs. More physically relevant simulations will likely require also some amounts of dissipation, dispersion, and/or inclusion of nonlinear effects. In linear situations, small dissipation, for example, as in

$$\frac{\partial u}{\partial t} = \frac{\partial u}{\partial x} + \varepsilon \frac{\partial^2 u}{\partial x^2} \tag{4.9}$$

with $0 < \varepsilon \ll 1$ will cause all Fourier modes in a solution to not just translate but also decay in magnitude. For the eigenvalues of a DM approximating the PDE RHS, eigenvalues that otherwise would be on the imaginary axis will be shifted to the left by frequency-dependent amounts. For many standard ODE solvers (leapfrog being an exception), they will still fall within their stability domains. Small amounts of dissipation tend to stabilize numerical time stepping. With large amounts, stability conditions that for explicit time stepping otherwise might have been of the form $k/h <$ const. are likely to become $k/h^2 <$ const.

Like (4.8) for pure convection, the heat equation

$$\frac{\partial u}{\partial t} = \frac{\partial^2 u}{\partial x^2} \tag{4.10}$$

is the simplest possible model for pure dissipation. It can nonetheless serve to illustrate the issue of *corner singularities*, which arise almost invariably at places where a boundary is nonsmooth, either in more than one space variable

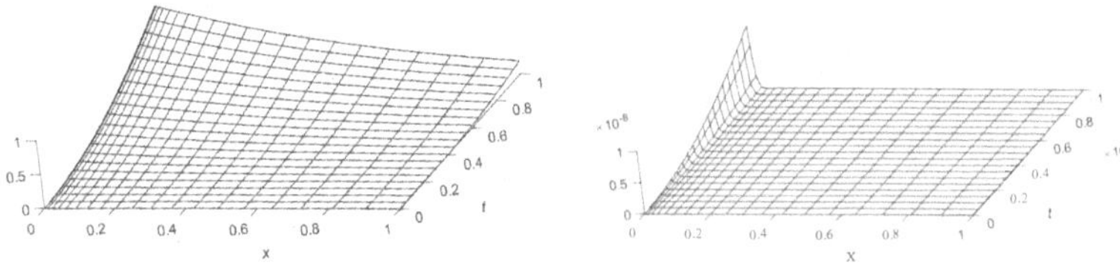

(a) Time interval $0 \leq t \leq 1$. (b) Time interval $0 \leq t \leq 0.0001$.

Figure 4.7 The heat equation corner solution (4.11) displayed over $x \in [0, 1]$ and two different time intervals.

or, as here, at a point where an initial condition (IC) in space meets a boundary condition (BC) in time.

As an example, one can directly verify that the analytic solution to (4.10) with IC $u(x, 0) = 0,\ x \geq 0$ and BC $u(t, 0) = t^2,\ t \geq 0$ is

$$u(x,t) = -\frac{x}{6}(10t + x^2)\sqrt{\frac{t}{\pi}}\, e^{-x^2/(4t)} + \left(t^2 + t\,x^2 + \frac{x^4}{12}\right)\operatorname{erfc}\left(\frac{x}{2\sqrt{t}}\right). \tag{4.11}$$

Figure 4.7(a) shows nothing surprising for this function. Nevertheless, keeping the x-interval displayed to $[0, 1]$ but in part (b) looking at a shorter initial interval in time reveals a boundary anomaly. This phenomenon is ubiquitous with initial-boundary value problems (Flyer and Fornberg, 2003) and is mostly invisible when using low-order approximation methods.[17] In the case of dispersive PDEs, such as $\frac{\partial u}{\partial t} = \frac{\partial^3 u}{\partial x^3}$, low-amplitude bursts of traveling waves can emanate from time-space corners.

4.2.3 Nonlinear Wave Equations

One of the most fundamental (and consequential) modern discoveries in applied mathematics was that of *solitons* (solitary waves emerging translated but otherwise not distorted after nonlinear interactions). These were first seen by Zabusky and Kruskal (1965) when they approximated solutions to the Korteweg–de Vries (KdV) equation

$$\frac{\partial u}{\partial t} + u\,\frac{\partial u}{\partial x} + \frac{\partial^3 u}{\partial x^3} = 0 \tag{4.12}$$

[17] Time-space corner singularities were first noted numerically with Chebyshev-based PS methods, which refine at boundaries for the different reason of suppressing Runge phenomenon-based artifacts (Boyd and Flyer, 1999).

using the FD2 scheme

$$\begin{aligned}&\frac{u(x,t+k)-u(x,t-k)}{2k}\\&\quad+\frac{u(x+h,t)+u(x,t)+u(x-h,t)}{3}\,\frac{(u(x+h,t)-u(x-h,t)}{2h}\\&\quad-\frac{u(x+2h,t)-2u(x+h,t)+2u(x-h,t)-u(x-2h,t)}{2h^3}=0.\end{aligned} \tag{4.13}$$

The soliton property was proved analytically shortly thereafter (Gardner et al., 1967). Zabusky and Kruskal also observed *nonlinear recurrences*. Figure 4.8 illustrates some such in the closely related modified KdV (or mKdV) equation

$$\frac{\partial u}{\partial t}+3u^2\,\frac{\partial u}{\partial x}+\frac{\partial^3 u}{\partial x^3}=0. \tag{4.14}$$

The initial condition $u(x,0) = 0.2\sin(7\pi/64)x$ features seven oscillations within the spatial period $x \in [0, 128]$. Due to the Benjamin–Feir instability (Benjamin and Feir, 1967), four spectral "side bands" dominated by the mode pairs $(6,8)$, $(5,9)$, $(4,1)$, and $(3,11)$ feature initially exponential growth. After starting from machine rounding errors, these modes initially grow exponentially, but decay again after reaching a certain size. The initial exponential growth rates can be calculated very accurately both for the analytic equation and for different spatial discretizations of it, providing the data for Figure 4.9.

As the number of nodes in the spatial period of this test problem is increased, the FD2-FD8 methods are seen to converge at the expected rates. The PS curve is more jagged (due to aliasing issues). Nevertheless, very coarse grids (n around 180) are seen to suffice for better than 10^{-10} accuracy and n = 256 (not displayed in the figure) suffices for around 10^{-15} precision. Similar extrapolation of the FD2 curve indicates that 10^{-15} precision would require $n \approx 2 \cdot 10^9$. Explicit time stepping schemes require $k/h^3 <$ const., making small h (large n) extraordinarily costly.[18]

The comparison in Figure 4.9 is typical for cases when the Fourier–PS method can be applied without facing complications related to boundaries and nontrivial domain geometries.

Consistent with the discussion above, nonlinear waves is an application area where very high order (in particular spectral and pseudospectral) methods have become routinely used, for example, Yang (2010), Chapter 7.

[18] If the cost is about 10^{-6} seconds per node each time step, total times would become well in excess of 10^{17} seconds – roughly the estimated age of the universe. In contrast, with periodic PS in space and a standard explicit ODE solver in time, a total computer time of around one second can be expected.

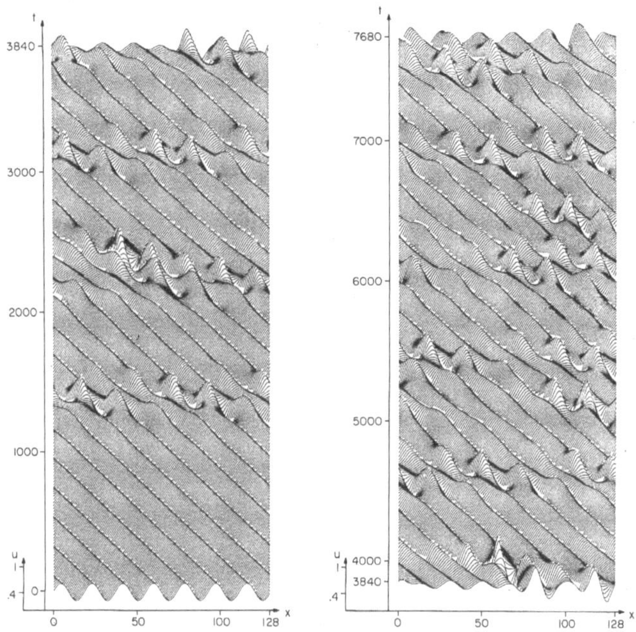

Figure 4.8 Fourier-PS solution of the mKdV equation (4.14) with a pure sine wave initial condition (bottom curve in left subplot; the subplot on the right starts at the end time in the left one). Figure reproduced from Fornberg and Whitham (1978).

4.2.4 Linearly Implicit Time Stepping Methods

Spatial derivatives of higher orders make the stability conditions for explicit time stepping methods more restrictive. However, a common feature of many nonlinear PDEs (such as the Navier–Stokes equations (5.15) and many nonlinear wave equations, e.g., (4.12), (4.14)) is that these highest derivatives in space occur only linearly and with constant coefficients. This provides opportunities for hybrid (linearly implicit) time stepping schemes, improving the computational efficiency relative to both purely explicit and purely implicit methods. The literature in this area is extensive, with a wide range of approaches, for example, Rosenbrock methods (related to implicit RK methods) (Rosenbrock, 1960), implicit-explicit (IMEX) methods (Ascher et al., 1997; Giraldo et al., 2013; Pareschi and Russo, 2005), IMEX enhanced by extrapolation (Cardone et al., 2014), exponential integrators (Hochbruck and Ostermann, 2010; Kassam and

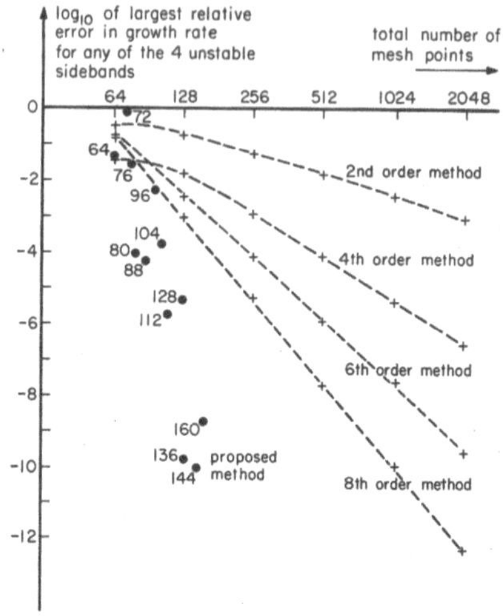

Figure 4.9 Accuracy of sideband growth rates for different spatial discretizations ("proposed method" refers to the PS method). Figure reproduced from Fornberg and Whitham, 1978.

Trefethen, 2005), integrating factor methods,[19] separation in frequency ranges ("sliders") (Fornberg and Driscoll, 1999), and time splitting (split step).[20] Several different methods for the nonlinear Schrödinger equation are compared Taha and Ablowitz (1984).

4.2.5 Nonlinear Conservation Laws

4.2.5.1 1-D Illustration of Conservation Law Concept

In its simplest scalar form in 1-D, an equation of this type could be

$$\frac{\partial u}{\partial t} + \frac{\partial f(u)}{\partial x} = 0, \tag{4.15}$$

where $f(u)$ is a nonlinear function of $u(x,t)$, say $f(u) = \frac{1}{2}u^2$. Mathematically, (4.15) can, in this case, alternatively be written as $u_t + u\, u_x = 0$, suggesting that u should stay constant along straight lines with slope u in the (x,t)-plane.

[19] Originally proposed for ODEs (Krusemeyer, 1994); some concerns in PDE contexts are raised in (Boyd, 2000), Chapter 14.

[20] Solving a time-dependent problem by alternating between simpler ones, some of which possibly allowing analytic solutions (Forest and Ruth, 1990; Lee and Fornberg, 2003; McLachlan and Quispel, 2002; Strang, 1968; Yoshida, 1990).

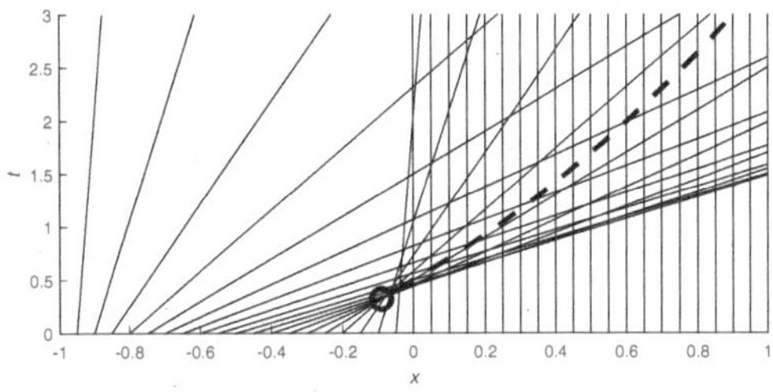

Figure 4.10 The (straight) characteristic lines in the case of (4.15) with initial condition (4.16). The circle and dashed curve mark the onset and later path of the shock discontinuity in the $\varepsilon \to 0$ limit of (4.17).

This is illustrated by the straight lines in Figure 4.10 for the initial condition

$$u(x,0) = \begin{cases} \sin(\pi x)^2 & -1 \le x < 0, \\ 0 & 0 \le x \le 1. \end{cases} \tag{4.16}$$

Intersecting lines represent apparent contradictions in a large area of the (x,t)-plane. However, equations such as (4.15) have usually arisen from physical principles (conservation laws) that also should have contained some very small diffusive term, such as

$$\frac{\partial u}{\partial t} + \frac{\partial f(u)}{\partial x} = \varepsilon \frac{\partial^2 u}{\partial x^2}, \tag{4.17}$$

with $0 < \varepsilon \ll 1$. Figures 4.11(a), (b) illustrate that the addition of the viscous term $\varepsilon \frac{\partial^2 u}{\partial x^2}$ makes the solutions single-valued[21] and, after a certain time, increasingly steep as ε is decreased. Since ε is extremely small in many applications, the limit of $\varepsilon \to 0$ is of particular interest. In this limit, a discontinuity (shock) arises at a certain instant in time and space[22] (circle marker in Figure 4.10), after which it travels as shown by the dashed curve.[23] Figure 4.12 shows the solution $u(x,t)$ at time $t = 2$. In the $\varepsilon \to 0$ limit, it is perfectly smooth on both sides of the discontinuity (at this time $t = 2$ located at $x \approx 0.604888$).

In many cases involving compressible fluids (and magnetohydrodynamics, etc.), qualitatively similar phenomena arise in 3-D and for systems of govern-

[21] A third derivative, as in (4.12) and (4.14), similarly ensures nonsingularity in many nonlinear wave contexts.

[22] In the present example at $t = \frac{1}{\pi}$, at location $x = -\frac{1}{4} + \frac{1}{2\pi}$.

[23] Velocities of these discontinuities depend on the sizes of the jumps in the dependent variables, as given by the *Rankine–Hugoniot* equations. In the present model equation, the speed is $\frac{1}{2}[u^2]/[u]$, where $[\,\cdot\,]$ denotes the jump in the respective variable. This can be seen by integrating (4.15) over an infinitesimal interval surrounding the discontinuity position.

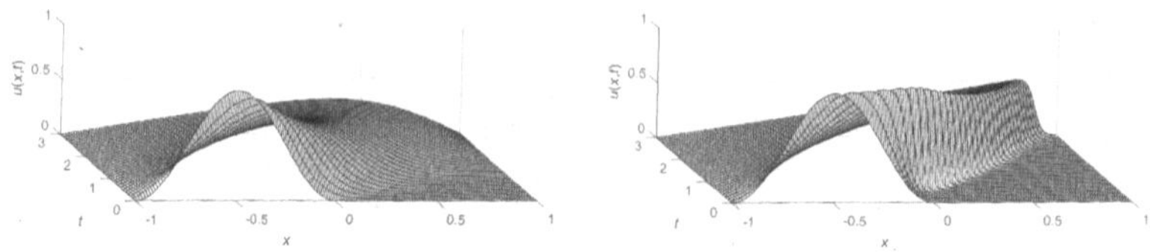

(a) $\varepsilon = 0.05$ (b) $\varepsilon = 0.005$

Figure 4.11 Numerical solutions to (4.17) with initial condition (4.16) and boundary conditions $u(\pm1, t) = 0$.

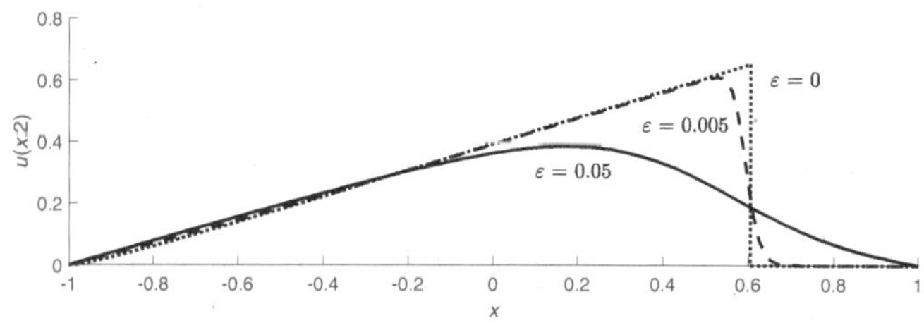

Figure 4.12 The solution to (4.17) with (4.16) and boundary conditions $u(\pm1, t) = 0$ at time $t = 2$ as $\varepsilon \to 0$.

ing equations, typically derived from conservation principles (such as mass, momentum, and energy).[24]

4.2.5.2 FD Solutions of Conservation Law Equations

There are numerous numerical challenges associated with solving conservation law equations, such as:

i. There can be other types of discontinuities (in function values or derivatives) present, known as contact discontinuities. These are not caused by characteristics crossing but are convected in a linear fashion, and are often best handled by different numerical techniques than shocks.
ii. Computing with ε small (or using some small artificial viscous counterpart) requires local refinement, which can get very costly especially in higher dimensions.
iii. Use of one-sided approximations require additional interface conditions (e.g., obtained by *Riemann solvers*), also complicated due to the variety

[24] For incompressible flows, one distinguishes between the Navier–Stokes and Euler equations, with the Reynolds number finite and infinite, corresponding to $\varepsilon > 0$ and $\varepsilon = 0$, respectively. In many such cases, there can be infinite families of $\varepsilon = 0$ solutions even when the $\varepsilon \to 0$ limit is unique (Fornberg and Elcrat, 2014).

of situations that can arise (e.g., collisions of curved shock surfaces). The Runge phenomenon is also often an issue with one-sided approximations.

Some low (first- or second-) order accurate FD schemes were developed around 1960 (e.g., Godunov's method and variations of the Lax–Wendroff method). However, to reach higher orders remains challenging in spite of vast efforts being given to the task.[25] Surveying these efforts falls outside the scope of the present book, and we refer to the monographs Gustafsson (2008); Hesthaven (2018); LeVeque (1990, 2002); Toro (2009).

4.3 Time-Independent PDEs

One way that time-independent PDEs arise is when time-dependent PDEs reach an equilibrium state (and with that, time derivatives vanish). Solving such PDEs via time stepping is often quite inefficient (although some solution methods can be interpreted as advancing in some suitably designed artificial time). More commonly, with no time present, the two key steps are: (i) Discretize to obtain an algebraic system of equations; and (ii) Solve this system effectively. Although these two parts are interlinked, we will follow the theme of this book and focus on (i), and in particular on how to improve the order of accuracy.

4.3.1 Poisson-Type PDEs

4.3.1.1 Compact Stencils in 2-D

The compact stencil concept was described in 1-D in Section 3.3.2. In higher-D, the range of PDEs for which this approach is practical is somewhat limited. Nevertheless, it is in certain cases both easy to implement and highly effective. The focus below is on Laplace, Poisson, and Helmholtz equations.[26]

Equations of Poisson and Laplace type arise in numerous situations, such as for a *stream function* in fluid mechanics or from *field equations* (describing for

[25] Recent developments for WENO (weighted essentially nonoscillatory methods) of up to around sixth order of accuracy are discussed in a special issue of *Communications on Applied Mathematics and Computation* (Vol. 5, Issue 1, 2023). Time stepping methods specialized to conservation law systems are surveyed in Gottlieb et al. (2011).

[26] Compact approximations have been developed also for certain other cases. Size 3×3 (9 point) fourth-order accurate approximations for the convection-diffusion equation $u_{xx} + u_{yy} + p(x, y)u_x + q(x, y)u_y = f(x, y)$ is described in Gupta et al. (1984) and for the steady 2-D Navier–Stokes equations in Li et al. (1995); Spotz and Carey (1995), with a 3-D generalization in Yu and Tian (2019). While the present description focuses on time-independent problems, compact stencils can also be applied in certain time-dependent situations (Lele, 1992; Spotz and Carey, 2001).

example gravitational and electrical fields). We consider as a first illustrative example the Poisson equation in 2-D

$$\frac{\partial^2 u}{\partial x^2} + \frac{\partial^2 u}{\partial y^2} = f\ . \tag{4.18}$$

Example 4.3.1 Create the following *compact* fourth-order accurate approximation

$$\begin{bmatrix} 1 & 4 & 1 \\ 4 & -20 & 4 \\ 1 & 4 & 1 \end{bmatrix} u/(6h^2) = \begin{bmatrix} & 1 & \\ 1 & 8 & 1 \\ & 1 & \end{bmatrix} f\ /12 + O(h^4) \tag{4.19}$$

for the 2-D Poisson's equation (a 2-D counterpart to the 1-D discretization in (1.7)).

Solution One way to arrive at (4.19) is to follow Collatz' *Mehrstellenverfahren* (Collatz, 1960; Fox, 1947). Because $\left(\frac{\partial^2}{\partial x^2} + \frac{\partial^2}{\partial y^2}\right) u = f$, it also holds that

$$\underbrace{\left(\frac{\partial^2}{\partial x^2} + \frac{\partial^2}{\partial y^2}\right)^2}_{\frac{\partial^4}{\partial x^4} + 2\frac{\partial^4}{\partial x^2 \partial y^2} + \frac{\partial^4}{\partial y^4}} u = \left(\frac{\partial^2}{\partial x^2} + \frac{\partial^2}{\partial y^2}\right) f. \tag{4.20}$$

Approximating (4.18) and (4.20) to fourth and to second order, respectively, gives

$$\begin{bmatrix} & & -\frac{1}{12} & & \\ & & \frac{4}{3} & & \\ -\frac{1}{12} & \frac{4}{3} & -5 & \frac{4}{3} & -\frac{1}{12} \\ & & \frac{4}{3} & & \\ & & -\frac{1}{12} & & \end{bmatrix} u/h^2 = [f] + O(h^4) \tag{4.21}$$

and

$$\begin{bmatrix} & & 1 & & \\ & 2 & -8 & 2 & \\ 1 & -8 & 20 & -8 & 1 \\ & 2 & -8 & 2 & \\ & & 1 & & \end{bmatrix} u/h^4 = \begin{bmatrix} & 1 & \\ 1 & -4 & 1 \\ & 1 & \end{bmatrix} f\ /h^2 + O(h^2)\ . \tag{4.22}$$

Adding $\frac{1}{12}h^2$ times (4.22) to (4.21) eliminates the outliers, and produces (4.19). □

As an approximation to the Laplace operator, the LHS of (4.19) is only accurate to second order, as seen by Taylor expanding it around the center point:

$$\begin{bmatrix} 1 & 4 & 1 \\ 4 & -20 & 4 \\ 1 & 4 & 1 \end{bmatrix} u/(6h^2)$$
$$= \left(\frac{\partial^2}{\partial x^2} + \frac{\partial^2}{\partial y^2}\right) u + \frac{1}{12} h^2 \left(\frac{\partial^2}{\partial x^2} + \frac{\partial^2}{\partial y^2}\right)^2 u$$
$$+ \frac{1}{360} h^4 \left(\frac{\partial^4}{\partial x^4} + 4\frac{\partial^4}{\partial x^2 \partial y^2} + \frac{\partial^4}{\partial y^4}\right)\left(\frac{\partial^2}{\partial x^2} + \frac{\partial^2}{\partial y^2}\right) u$$
$$+ \frac{1}{60480} h^6 \left(\frac{\partial^4}{\partial x^4} + 4\frac{\partial^4}{\partial x^2 \partial y^2} + \frac{\partial^4}{\partial y^4}\right)\left(3\frac{\partial^4}{\partial x^4} + 16\frac{\partial^4}{\partial x^2 \partial y^2} + 3\frac{\partial^4}{\partial y^4}\right) u$$
$$+ O(h^8). \tag{4.23}$$

When approximating (4.18), note that $\left(\frac{\partial^2}{\partial x^2} + \frac{\partial^2}{\partial y^2}\right)^2 u = \left(\frac{\partial^2}{\partial x^2} + \frac{\partial^2}{\partial y^2}\right) f$. The second term in the RHS of (4.23) together with the standard 5-point approximation for $\left(\frac{\partial^2}{\partial x^2} + \frac{\partial^2}{\partial y^2}\right) f$ provides an alternate way to verify (4.19). For solutions to Laplace's equation $\left(\frac{\partial^2}{\partial x^2} + \frac{\partial^2}{\partial y^2}\right) u = 0$, also the third RHS terms vanishes, and the approximation becomes sixth-order accurate.[27] Key advantages of (4.19) over (4.21) include:

- the compact stencil is easier to use near boundaries;
- the diagonal dominance of coefficient matrix improves numerical stability and speeds up many iterative solution methods;
- for the same order of accuracy, error constants when approximating derivatives become smaller when a stencil uses more local and less far away information.

Variations on the idea of designing compact FD approximations for elliptic PDEs that utilize analytic solution features can be found in Gupta (1991); Lynch and Rice (1978); Manohar and Stephenson (1984).

[27] An alternative way to obtain this result is given in Section 6.4 (not depending on the "lucky" factorization in the third term in the RHS of (4.23)).

4.3.1.2 Compact Stencils in 3-D

Regarding $3 \times 3 \times 3$ size FD stencils approximating $\Delta u = \frac{\partial^2 u}{\partial x^2} + \frac{\partial^2 u}{\partial y^2} + \frac{\partial^2 u}{\partial z^2} = f$, series expansion as in (4.23) shows that

$$\left[\begin{array}{ccc} 1 & 3 & 1 \\ 3 & 14 & 3 \\ 1 & 3 & 1 \\ \hline 3 & 14 & 3 \\ 14 & -128 & 14 \\ 3 & 14 & 3 \\ \hline 1 & 3 & 1 \\ 3 & 14 & 3 \\ 1 & 3 & 1 \end{array}\right] u/(30h^2) = \left[\begin{array}{ccc} 0 & 0 & 0 \\ 0 & 1 & 0 \\ 0 & 0 & 0 \\ \hline 0 & 1 & 0 \\ 1 & 6 & 1 \\ 0 & 1 & 0 \\ \hline 0 & 0 & 0 \\ 0 & 1 & 0 \\ 0 & 0 & 0 \end{array}\right] f/12 + O(h^4) \qquad (4.24)$$

also becomes accurate to $O(h^6)$ when applied to solutions of Laplace's equation (the $f(x, y, z) \equiv 0$ case).[28] Additional discussion on the 3-D Poisson case can be found in Sutmann and Steffen (2006).

Similar to the situation in 2-D, the LHS of (4.24) is by itself only a second-order accurate approximation to the Laplacian operator $\Delta = \frac{\partial^2}{\partial x^2} + \frac{\partial^2}{\partial y^2} + \frac{\partial^2}{\partial z^2}$, but the residual when applying the stencil (omitting the factor $1/(30h^2)$) to a Laplace equation solution will be $O(h^8)$.

4.3.2 Laplace's Equation in 3-D

Harmonic functions (solutions to the Laplace equation) arise in a large number of applications, for example, in connections with gravitational and electrical fields. In 2-D, a harmonic function can be viewed as the real (or imaginary) part of an analytic function, and complex variable techniques may apply, see Chapter 6. We focus next on FD formulas that are specific to harmonic functions in 3-D (satisfying $\Delta u \equiv \left(\frac{\partial^2}{\partial x^2} + \frac{\partial^2}{\partial y^2} + \frac{\partial^2}{\partial z^2}\right) u = 0$), for which no similar option is available.

As noted earlier, the $3 \times 3 \times 3$ stencil in the LHS of (4.24), if omitting the factor $u/(30h^2)$, will evaluate to $O(h^8)$ if applied to a harmonic function. Its $5 \times 5 \times 5$ counterpart is shown in (4.25), this time normalized to have the central element one and a remaining free parameter used to set the cube corner elements to zero. It has very small coefficients outside its central $3 \times 3 \times 3$

[28] Extending to 3-D the approach in Example 4.3.1 gives different u-weights than shown in (4.24) These also give $O(h^4)$ accuracy for the Poisson equation, but do not improve to $O(h^6)$ for Laplace's equation.

region, but nevertheless reduces the residual to $O(h^{16})$.[29]

$$\left[\begin{array}{ccccc} m & j & g & j & m \\ j & i & f & i & j \\ g & f & e & f & g \\ j & i & f & i & j \\ m & j & g & j & m \\ \hline j & i & f & i & j \\ i & d & c & d & i \\ f & c & b & c & f \\ i & d & c & d & i \\ j & i & f & i & j \\ \hline g & f & e & f & g \\ f & c & b & c & f \\ e & b & a & b & e \\ f & c & b & c & f \\ g & f & e & f & g \\ \hline j & i & f & i & j \\ i & d & c & d & i \\ f & c & b & c & f \\ i & d & c & d & i \\ j & i & f & i & j \\ \hline m & j & g & j & m \\ j & i & f & i & j \\ g & f & e & f & g \\ j & i & f & i & j \\ m & j & g & j & m \end{array}\right], \quad \left\{\begin{array}{llll} a = & 1, & & \\ b & -\dfrac{410242}{3444909} & \approx & -0.119086454823625, \\ c = & -\dfrac{215911}{10334727} & \approx & -0.020891795206588, \\ d = & -\dfrac{6657}{1148303} & \approx & -0.005797250377296, \\ e = & \dfrac{3645}{2296606} & \approx & 0.001587124652640, \\ f = & \dfrac{568}{3444909} & \approx & 0.000164880988148, \\ g = & \dfrac{707}{20669454} & \approx & 0.000034205064149, \\ i = & -\dfrac{1993}{20669454} & \approx & -0.000096422479278, \\ j = & \dfrac{1}{1087866} & \approx & 0.000000919230861, \\ m = & 0. & & \end{array}\right. \tag{4.25}$$

In the counterpart expansion to (4.23) (again, due to symmetries, containing only even powers of h), the choice given above of weights $a, b, \ldots, j, m$ ensures that every term up through $O(h^{14})$ for the stencil in (4.25) contains the factor $\left(\frac{\partial^2}{\partial x^2} + \frac{\partial^2}{\partial y^2} + \frac{\partial^2}{\partial z^2}\right) u$. Therefore, all these terms vanish for solutions to Laplace's equation.[30]

Example 4.3.2 For a 3-D harmonic function, numerically compare the convergence rates as the grid spacing is reduced when applying the standard 7-point approximation (3-D counterpart to $\left[\begin{smallmatrix} & 1 & \\ 1 & -4 & 1 \\ & 1 & \end{smallmatrix}\right] / h^2$ in 2-D) and the $3 \times 3 \times 3$ and $5 \times 5 \times 5$ compact stencils described earlier.

Solution Applying the respective stencils, normalized to have the central weight equal to 1 and centered at the origin, to the function $f(x, y, z) = e^{5z} \sin(3(x + \frac{1}{2})) \cos(4y)$ gives the residual ("Error") shown in Figure 4.13.

[29] Extending to $7 \times 7 \times 7$ stencil size, there is a unique set of weights that brings the residual to $O(h^{26})$.

[30] Expanding in h, as for (4.23) but with undetermined stencil weights $a, b, c, \ldots$ and enforcing $\left(\frac{\partial^2}{\partial x^2} + \frac{\partial^2}{\partial y^2} + \frac{\partial^2}{\partial z^2}\right)$ to be a factor for successive powers of h, gives rise to linear equations for the weights.

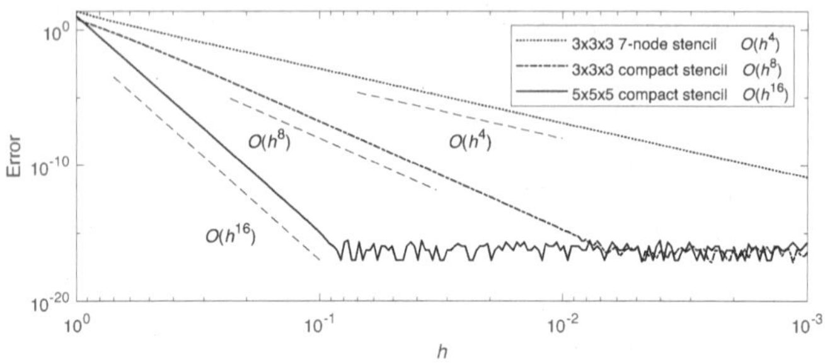

Figure 4.13 Convergence rates in the test described in Example 4.3.2.

The convergence rates for decreasing h precisely match the theoretically predicted ones of $O(h^4)$, $O(h^8)$, and $O(h^{16})$, respectively. The 7-node stencil is grossly inefficient. The $5 \times 5 \times 5$ stencil is seen to provide machine precision accuracy already with a 10 times larger grid spacing h than its $3 \times 3 \times 3$ counterpart. Since this is in 3-D, this ratio corresponds to needing only $1/1000$ the number of grid points compared to what would be needed with the compact $3 \times 3 \times 3$ stencil. □

4.3.3 FD Approximations for Interpolation and Derivatives of 3-D Harmonic Functions

If it is known that a function is harmonic, this allows for an alternative to typical tensor-style approximations (using for each derivative only data along a 1-D grid line). The key idea is to note that, when a general 3-D function is Taylor expanded around a stencil center point (say, the origin) in powers in x, y, z up through some degree d, this requires $\binom{d+3}{3} = \frac{1}{6}(d+3)(d+2)(d+1)$ monomial terms. When knowing that a 3-D function is harmonic, the similar expansion requires only $(d+1)^2$ terms (corresponding to all volume spherical harmonic functions up through degree d). An FD formula needs about the same number of nodes as there are expansion terms. With much fewer terms to consider in the harmonic case, relatively small 3-D stencils can then become attractive alternatives to traditional 1-D approximations (which are unable to exploit any special features of harmonic functions).[31]

[31] There are some conceptual similarities between this approach and Trefftz methods , as used for example to approximate solutions to the Helmholtz equation (Hiptmair et al., 2016; Lambe et al., 2003). In both cases, a key aspect is to limit the function space that the approximations are based on to either an approximate or the full subspace of solutions of the target PDE.

Example 4.3.3 For a function that satisfies the 3-D Laplace equation, interpolate to a half-way point along an equispaced Cartesian grid line.

Solution Simple formulas for this type of interpolation are convenient in different contexts, such as when refining from a 3-D grid to one with halved grid spacing. For a general function (not known to satisfy Laplace's equation), 1-D interpolation along the grid line is a natural option. Table 4.3 gives the leading order FD stencil weights for this case. With the additional knowledge that the function to be interpolated satisfies Laplace equation, we can alternatively seek weights in a similar way as for the Laplace operator case (4.25). This gives, for example, the 10th-order accurate weight set (interpolating in the z-direction):

$$\left[\begin{array}{ccccc}
0 & 0 & 0 & 0 & 0 \\
0 & j & i & j & 0 \\
0 & i & g & i & 0 \\
0 & j & i & j & 0 \\
0 & 0 & 0 & 0 & 0 \\ \hline
f & e & d & e & f \\
e & c & b & c & e \\
d & b & a & b & d \\
e & c & b & c & e \\
f & e & d & e & f \\ \hline
f & e & d & e & f \\
e & c & b & c & e \\
d & b & a & b & d \\
e & c & b & c & e \\
f & e & d & e & f \\ \hline
0 & 0 & 0 & 0 & 0 \\
0 & j & i & j & 0 \\
0 & i & g & i & 0 \\
0 & j & i & j & 0 \\
0 & 0 & 0 & 0 & 0
\end{array}\right] u \qquad
\begin{cases}
a = \dfrac{3463}{81920} & \approx \quad 0.042272949218750, \\
b = \dfrac{381}{5120} & \approx \quad 0.074414062500000, \\
c = \dfrac{1487}{81920} & \approx \quad 0.018151855468750, \\
d = \dfrac{37}{32768} & \approx \quad 0.001129150390625, \\
e = \dfrac{7}{8192} & \approx \quad 0.000854492187500, \\
f = \dfrac{3}{65536} & \approx \quad 0.000045776367188, \\
g = \dfrac{81}{2560} & \approx \quad 0.031640625000000, \\
i = \dfrac{161}{20480} & \approx \quad 0.007861328125000, \\
j = \dfrac{263}{81920} & \approx \quad 0.003210449218750.
\end{cases} \tag{4.26}$$

These weights are all positive, and decay much faster with the distance from the interpolation point than is the case for unidirectional FD approximations (as seen in Table 4.3), resulting in an accuracy advantage even in cases when the formal accuracy orders are the same.[32] □

Rapid decay of weights with the distance from a stencil's center is a typical feature when FD approximations are derived specifically for harmonic func-

[32] Less wide stencils are also more convenient near boundaries.

Table 4.3 *Weights for interpolation to a half-way point on an equispaced 1-D grid, accurate at the stencils' midpoints.*

Order	Weights interpolation										
2					$\frac{1}{2}$	$\frac{1}{2}$					
4				$-\frac{1}{16}$	$\frac{9}{16}$	$\frac{9}{16}$	$-\frac{1}{16}$				
6			$\frac{3}{256}$	$-\frac{25}{256}$	$\frac{75}{128}$	$\frac{75}{128}$	$-\frac{25}{256}$	$\frac{3}{256}$			
8		$-\frac{5}{2048}$	$\frac{49}{2048}$	$-\frac{245}{2048}$	$\frac{1225}{2048}$	$\frac{1225}{2048}$	$-\frac{245}{2048}$	$\frac{49}{2048}$	$-\frac{5}{2048}$		
10	$\frac{35}{65536}$	$-\frac{405}{65536}$	$\frac{567}{16384}$	$-\frac{2205}{16384}$	$\frac{19845}{32768}$	$\frac{19845}{32768}$	$-\frac{2205}{16384}$	$\frac{567}{16384}$	$-\frac{405}{65536}$	$\frac{35}{65536}$	
$\vdots$	$\downarrow$	$\downarrow$	$\downarrow$	$\downarrow$	$\downarrow$	$\downarrow$	$\downarrow$	$\downarrow$	$\downarrow$	$\downarrow$	
Limit $\cdots$	$\frac{2}{9\pi}$	$-\frac{2}{7\pi}$	$\frac{2}{5\pi}$	$-\frac{2}{3\pi}$	$\frac{2}{1\pi}$	$\frac{2}{1\pi}$	$-\frac{2}{3\pi}$	$\frac{2}{5\pi}$	$-\frac{2}{7\pi}$	$\frac{2}{9\pi}$	$\cdots$

tions. This aspect has been studied more extensively in the 2-D case, see Section 6.3.1. The counterpart problem in 2-D to Example 4.3.3 was studied in Section 7 of Fornberg (2022). In that case, particularly effective formulas can be found if the target function is analytic (with both real and imaginary parts available).

4.3.4 Compact Stencils for the 3-D Helmholtz Equation

The Helmholtz equation typically arises as an eigenvalue problem

$$\left(\frac{\partial^2}{\partial x^2} + \frac{\partial^2}{\partial y^2} + \frac{\partial^2}{\partial z^2}\right) u = \lambda u. \tag{4.27}$$

With all its eigenvalues negative, another common notation is to write $\lambda = -k^2$. The equation is of interest also for λ positive and is then known as the Modified Helmholtz equation. The treatment of the 3-D Laplace equation in Sections 4.3.2 and 4.3.3 was partly motivated by the fact that the approach used there generalizes quite directly to 3-D Helmholtz equations, as illustrated in the following three examples.

Example 4.3.4 The LHS stencil in (4.24), omitting the division by h^2, was in Example 4.3.2, seen to evaluate to $O(h^8)$ for solutions to the 3-D Laplace's equation. Give a counterpart formula for to 3-D Helmholtz solutions.

Solution Consider the h-spaced 3-D FD stencil

$$H\,u = \left[\begin{array}{ccc} d & c & d \\ c & b & c \\ d & c & d \\ \hline c & b & c \\ b & a & b \\ c & b & c \\ \hline d & c & d \\ c & b & c \\ d & c & d \end{array}\right] u \text{ with } \left\{\begin{array}{lcllll} a & = & 1, & & & \\ b & = & -\frac{7}{64} & +\frac{13}{192}(\lambda h^2) & -\frac{1}{512}(\lambda h^2)^2, \\ c & = & -\frac{3}{128} & -\frac{19}{768}(\lambda h^2), & \\ d & = & -\frac{1}{128} & +\frac{1}{64}(\lambda h^2). & \end{array}\right. \tag{4.28}$$

For the parameter value $\lambda = 0$, this reduces to a scaled version of the LHS of (4.24). Direct series expansion of Hu around the stencil center point will show that the coefficients for h^0 and for all odd powers of h vanish, and the coefficients for h^2, h^4, h^6 all contain the factor $\left(\frac{\partial^2}{\partial x^2} + \frac{\partial^2}{\partial y^2} + \frac{\partial^2}{\partial z^2} - \lambda\right) u$. With the expansions for a, b, c, d truncated as shown, the stencil weight set is unique. □

Example 4.3.5 Compare numerically the residual when we to a known Helmholtz solution apply the compact stencil (4.28) versus the standard FD2 approximation, as obtained by the weight choice $a = 1 - \frac{1}{6}(\lambda h^2)$, $b = -\frac{1}{6}$, $c = d = 0$.

Solution We choose quite arbitrarily as test function $u(x, y, z) = \sin(2.3(x - 0.3))\sin(3.5(x - 0.7))\sin(-1.2(x - 1.1))$ for which $\lambda = -18.98$ makes $\Delta u - \lambda u \equiv 0$, and apply to it at the origin (i) the stencil as given in the example text (FD2) and (ii) the compact stencil (4.28). Figure 4.14 shows the resulting residuals. If one wishes to approximate Laplace or Helmholtz solutions, there would also be a division by h^2 (reducing the slopes of the initial parts of the two curves, and causing the rounding errors for small h to rise like $O(10^{-16}/h^2)$). □

Example 4.3.6 Approximate at the center of $3 \times 3 \times 3$ stencils $\frac{\partial u}{\partial z}$ and $\frac{\partial^2 u}{\partial z^2}$ for Helmholtz solutions $u(x, y, z)$.

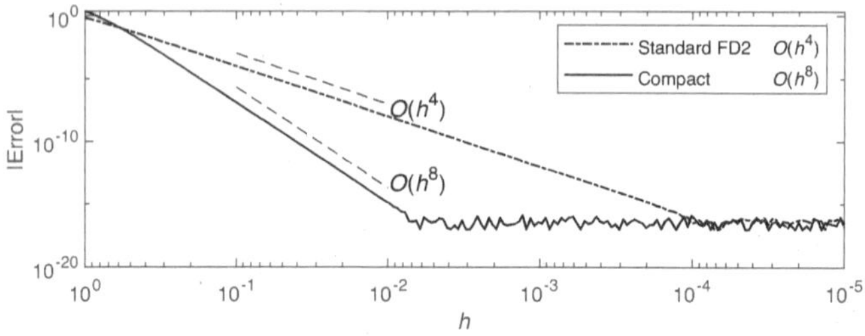

Figure 4.14 The residuals in the test described in Example 4.3.5. When interpreting this figure (and previously Figure 4.13) we recall that, for a volume discretization, the total number of node points scales as $O(1/h^3)$.

Solution Direct series expansions show that

$$\frac{\partial u}{\partial z} = \left[\begin{array}{ccc} 0 & 1 & 0 \\ 1 & 2 & 1 \\ 0 & 1 & 0 \\ \hline 0 & 0 & 0 \\ 0 & 0 & 0 \\ 0 & 0 & 0 \\ \hline 0 & -1 & 0 \\ -1 & -2 & -1 \\ 0 & -1 & 0 \end{array}\right] \frac{u}{12h}\left(1 - \frac{1}{6}\lambda h^2\right) - \frac{h^2}{6}\frac{\partial}{\partial z}\left(\frac{\partial^2}{\partial x^2} + \frac{\partial^2}{\partial y^2} + \frac{\partial^2}{\partial z^2} - \lambda\right) u + O(h^4)$$

and

$$\frac{\partial^2 u}{\partial z^2} = \left[\begin{array}{ccc} 0 & 1 & 0 \\ 1 & 8 & 1 \\ 0 & 1 & 0 \\ \hline 0 & -2 & 0 \\ -2 & -16 & -2 \\ 0 & -2 & 0 \\ \hline 0 & 1 & 0 \\ 1 & 8 & 1 \\ 0 & 1 & 0 \end{array}\right] \frac{u}{12h^2}\left(1 - \frac{1}{12}\lambda h^2\right) - \frac{h^2}{12}\frac{\partial^2}{\partial z^2}\left(\frac{\partial^2}{\partial x^2} + \frac{\partial^2}{\partial y^2} + \frac{\partial^2}{\partial z^2} - \lambda\right) u + O(h^4).$$

The special case of $\lambda = 0$ give corresponding formulas for Laplace solutions. □

4.3.5 Brief Comments on Direct vs. Iterative Solvers

FD-discretized time-independent PDEs typically lead to large and sparse linear systems of equations, as illustrated in the ODE BVP case in (3.14)) and, after linearization by Newton's method, in (3.17). This brings up a choice between direct and iterative linear solvers. Both these categories of solvers are described

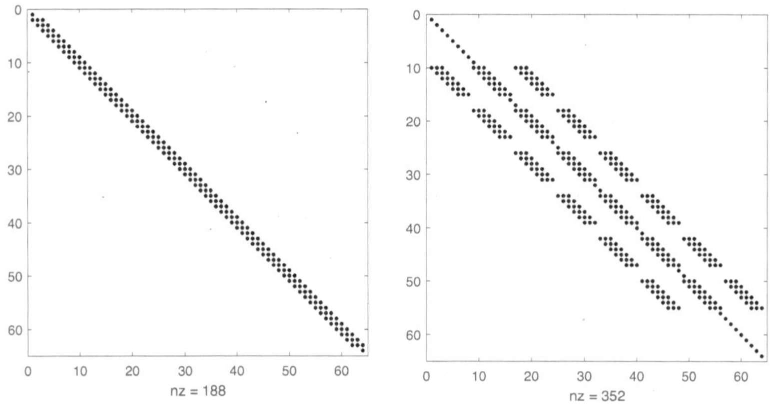

(a) DM in 1-D (64 nodes) (b) DM in 2-D (nodes on an 8 × 8 grid)

Figure 4.15 Sparsity patterns for DMs approximating the Laplacian operator (together with Dirichlet boundary condition equations).

extensively in the literature. As these fall outside the scope of this book, we limit ourselves here to a few general remarks.

Direct solvers: These algorithms are mathematically capable of producing the exact result in a finite number of steps. The first step is usually to form a DM, in the style shown in the ODE cases cited in (3.14) and (3.17). Figure 4.15 illustrates how the DM pattern of nonzero entries changes from 1-D to 2-D in the case of approximating the Laplacian (using the stencils $\begin{bmatrix} 1 & -2 & 1 \end{bmatrix}/h^2$ and $\begin{bmatrix} 1 & 4 & 1 \\ 4 & -20 & 4 \\ 1 & 4 & 1 \end{bmatrix}/(6h^2)$, respectively). In both cases shown, there are 64 nodes in total, that is, in the 2-D case placed on an 8×8 grid. The pattern in part (a) matches that in (3.14). In case there are n nodes in each spatial direction, the cost of solving with pivoted Gaussian elimination (GE) grows severely with the number of dimensions, as $O(n)$, $O(n^4)$, and $O(n^7)$ in 1-D, 2-D, and 3-D, respectively.[33] One situation that can offer some advantage for direct solvers occurs if many cases have to be solved, differing only in their right-hand sides (RHSs). It can then pay off to spend extra effort on achieving sparse matrix factorizations, which can be re-used repeatedly (possibly in combination with finding an approximate low-rank matrix decomposition: see Halko et al., 2011).

[33] The cost of solving a banded linear system by pivoted Gaussian elimination scales proportionally to the number of equations and the square of the bandwidth.

Iterative solvers: These solvers bear some conceptual resemblance to stepping forward time-dependent PDEs, but in an artificial time that is designed to accelerate (and stabilize) convergence. In general, these methods become increasingly advantageous as the DMs become larger and sparser (as is typical in cases of several spatial dimensions). In many cases, the DMs need not be formed explicitly. Many software libraries provide a range of excellent iterative solvers, often with the main issue for the user being to find a good preconditioner. The effectiveness of most iterative solvers depends on the details of the system to be solved. For example, for (4.27), most iterative solvers encounter difficulties if $\lambda < 0$ (Ernst and Gander, 2012).[34]

If a linear system matrix is known to be positive definite, the conjugate gradient method can be applied. Although technically a direct solver, it is virtually always used as an iterative one. The method was first described in Hestens and Stiefel (1952). For a survey both of the original method and of its more recent variations (including for nonsymmetric matrices), see for example Golub and Loan (2013).

When applicable, multigrid (MG) methods can provide convergence in a number of operations that scales proportionally to the total number of unknowns (rows in the DM, if such is formed)[35] (Brandt, 1977; Briggs et al., 2000). While there are many versions available, the two main concepts are Geometric and Algebraic (with abbreviations GMG and AMG, respectively). Underlying these is the concept of iterative improvement, see Example 3.3.1.

Geometric Multigrid: These require a hierarchy of node set and converge by employing two complementary processes: smoothing and coarse-grid corrections. Smoothing (relaxation) can be carried out by a simple iterative method, such as Gauss–Seidel or Jacobi. Information is then transferred to a coarser grid through *restriction*. On the coarser grid, the problem is either solved directly or the residual is smoothed, followed by moving to still coarser grid. Transferring back to the finer grids involves interpolation (*prolongation*). The solution process requires cycling up and down between grids until all residuals fall within tolerances. The basic concept is that smoothing reduces high-frequency errors at each grid level, and thereby at all levels. Strengths include relatively easy implementation and intuitive geometric interpretation.

Algebraic Multigrid: These methods solve linear systems using multigrid principles, but in ways that depend only on the coefficients of a (typically

[34] Note, however, the iterative scheme described in Bayliss et al. (1983). If both λ and u are to be determined, the problem is nonlinear, and the Newton-based scheme (Peters and Wilkinson, 1979) is applicable (however requiring a linear solve within each iteration).

[35] MG can sometimes also be used as preconditioners for other iterative methods.

sparse) matrix, with no direct link to an underlying PDE or discretization method. They decompose the operator rather than build node sets with an intuitive geometric interpretation (therefore they are also referred to as algebraic multilevel methods). AMG determines levels, inter-level transfer operators, and coarse-level operators based on the matrix entries. The different levels are nested subsets of unknowns which, like inter-grid transfer operators, have no intuitive geometric meaning. AMG is surveyed, for example, in Falgout (2006); Shapira (2008); Stüben (2002). Strengths of AMG include no requirement of knowledge of problem or geometry, and often simplicity in parallelization. When both GMG and AMG are applicable, the former may be advantageous in terms of computational speed.

4.4 Irregularly Shaped Boundaries

If it was not for boundaries, the subject of numerical treatment of PDEs would have been much simpler. The key choice when boundaries do not align with grid lines is whether to still use a grid-based discretization or to change to a mesh-free method. We make some remarks about the former here, leaving the latter to Chapter 5.

There exists an extensive literature on FD approximations in connection with arbitrarily shaped boundaries/interfaces, that are either fixed in position or moving (then often according to the evolution of the PDE solution). A related problem category is time-independent free boundary problems. Since grid-based FD methods in these problem categories for their accuracy often rely more on grid refinements than on higher orders of accuracy, we limit ourselves here to naming a few types of approaches, together with some key references.

Immersed boundaries/interfaces: Curved boundaries cut across the grid, and FD weights in affected stencils are modified to compensate for this (without altering grid density) (Balam and Zapata, 2022; Li and Ito, 2006; Mittal and Iaccarino, 2005; Peskin, 1977, 2002).[36]

Adaptive mesh refinement: With all grid lines remaining in the directions of the coordinate axes, the grid spacing is repeatedly halved in the vicinity of boundaries and interfaces. In time-dependent situations, refinement can be either added or removed (Berger and Colella, 1989; Berger et al., 2011; Berger and Oliger, 1984).

[36] Low-order accurate corrections to the basic 5-point stencil for the 2-D Laplacian with a fixed curved boundary are described in Morton and Mayers (1994), Section 3.4, with a free boundary in Fornberg (2010), and for the 2-D streamfunction-vorticity equations in Calhoun (2002).

Overlapped (overset) grids: Rectangular (in 2-D) grid patches are by transformations in the independent variables changed to locally fit boundaries/interfaces, and these patches can then be connected by means of interpolation (Chesshire and Henshaw, 1990; Henshaw and Schwendeman, 2006). Different coupling methods are compared in Driscoll and Fornberg (1999) for 2-D discontinuous coefficient Maxwell's equations. A combination of overset grids with stabilizing upwinding is discussed in Angel et al. (2023).

Level-set methods: Level-set methods can be highly effective, especially for time-evolving boundaries that involve changes in topology (such as domains splitting in pieces, developing holes, etc.). In such cases, these permit computations to be carried out also on fixed Cartesian grids. For a recent review, see Gibou et al. (2018).

While the presence of interfaces and boundaries greatly increases code complexity in most cases, the increase in computational cost can still be quite low, as these occur in one dimension lower than that of the full domain (at points in 1-D, along curves in 2-D, and over surfaces in 3-D).

4.5 Infinite Domains

One option in the case of infinite domains is to make the domain bounded by means of variable change(s) before discretization – typically at the expense of introducing singularities in the governing PDEs. Another (more commonly used) option is to create boundary conditions (BCs) that are compatible with far-field asymptotics. This is most simply done in time-independent cases.[37] However, the need arises particularly frequently in applications with wave equations (acoustic, elastic, Maxwell's equations, etc.), when outgoing waves ideally should propagate undisturbed to infinity.[38]

A breakthrough on creating nonreflecting boundary conditions was achieved by the introduction of perfectly matched layers (PML) in Berenger (1994), designed in such way that waves entering the layer get absorbed without any reflections. Following the initial discovery that this concept was possible, numerous variations of it have been developed, for example, Appelö et al. (2006); Appelö and Kreiss (2006); Berenger (2022); Fathi et al. (2015); Johnson (2021); Oskooi and Johnson (2011).

[37] One such example, for 3-D steady axisymmetric flow past a sphere, is given in Fornberg (1988b), utilizing the fact that outside a thin vortex region, the Navier–Stokes equations simplify to the 3-D Laplace equation, with solutions of interest decaying outwards.

[38] Variable changes to bring infinity to a finite distance is not an option (as the wave length would decrease to zero, and the waves would not be numerically resolvable). Much better options are described in Bayliss and Turkel (1980); Engquist and Majda (1979).

5

Mesh-Free FD Approximations

Chapter 2 described how local FD approximations lead to global PS methods when their stencil sizes are increased. While the accuracy increased from algebraic to spectral, this comes in many applications also with some drawbacks:

i. Very reduced geometric flexibility (even in the Chebyshev-type nonperiodic case, as that primarily is applicable only on finite intervals in each space direction).
ii. Limited opportunities for local refinements.
iii. Loss of spatial locality of derivative approximations.

The need for methods that address these issues, as well as providing high accuracy orders at low computational cost, was recognized around the same time as when PS methods first emerged (1972). This was then expressed in Jensen (1972) as:

> The ideal is to have a "node generator" which could sprinkle nodes on the problem domain with a density varying according to some given "density function". This density function should then be an approximation to the truncation error in the finite difference expressions used, which of course depends on the solution.

This was accompanied with the conceptual illustration reproduced in Figure 5.1. As described in Section 5.1, finding weights for scattered node FD stencils is not merely a matter of differentiating multivariate interpolating polynomials. Two main approaches for "generalized FD methods" are known as Moving Least Squares (MLS) and Radial Basis Function-generated Finite Differences (RBF–FD).[1] Section 5.2 briefly describes the concept of MLS, Section 5.3 introduces RBFs, and Section 5.4 describes the spatially localized RBF–FD approach. There are only few direct MLS versus RBF–FD comparisons in the

[1] Some earlier methods are described in Liszka and Orkisz (1980), in practice often limited to second- or third-order accuracy Gavete et al. (2003); Zheng and Li (2022).

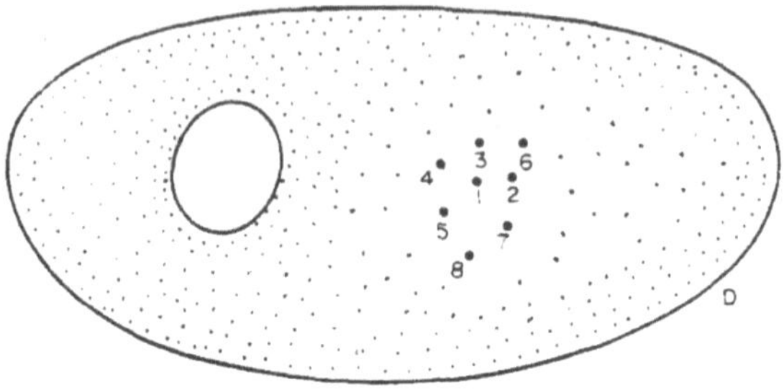

Figure 5.1 Schematic illustration of a scattered node FD stencil (reproduced with permission from Jensen (1972)).

literature. It is argued in Bayona (2019a) that RBF–FD tends to be advantageous especially in cases of high accuracy requirements.

5.1 Difficulty with Taylor Expansion-Based FD on Scattered Nodes

The algorithm in Section 1.2.1.1 provided one (of several) ways to obtain the weights in a regular (polynomial-based) FD approximation in 1-D. We showed there that the coefficient matrix that arose was nonsingular for distinct nodes. This matrix is the transpose of the one that arises when one seeks an interpolating polynomial $p(x) = a_1 + a_2x + \cdots + a_nx^{n-1}$ for the data $f(x_k) = f_k$, $k = 1, 2, \ldots, n$, by directly enforcing the interpolation conditions, which thus also is nonsingular.[2] Denoting the successive monomials $1, x, x^2, \ldots, x^{n-1}$ by $F_1(x), F_2(x), F_3(x), \ldots, F_n(x)$, the system for obtaining polynomial interpolation coefficients becomes

$$\begin{bmatrix} F_1(x_1) & F_2(x_1) & \cdots & F_n(x_1) \\ F_1(x_2) & F_2(x_2) & \cdots & F_n(x_2) \\ \vdots & \vdots & & \vdots \\ F_1(x_n) & F_2(x_n) & \cdots & F_n(x_n) \end{bmatrix} \begin{bmatrix} a_1 \\ a_2 \\ \vdots \\ a_n \end{bmatrix} = \begin{bmatrix} f_1 \\ f_2 \\ \vdots \\ f_n \end{bmatrix}. \tag{5.1}$$

This nonsingularity in 1-D is no longer assured in 2-D and above, *no matter how the basis functions $F_k(\underline{x})$ are chosen* (polynomial, or not). For example, in 2-D, let $\underline{x}$ denote an arbitrary position $\underline{x} = \left[\begin{smallmatrix} x \\ y \end{smallmatrix}\right]$ and let the node locations be

[2] See method iv in Appendix A.

$\underline{x}_k = \left[\begin{smallmatrix} x_k \\ y_k \end{smallmatrix}\right]$, $k = 1, 2, \ldots, n$. The difference to (5.1) may appear minimal:

$$\begin{bmatrix} F_1(\underline{x}_1) & F_2(\underline{x}_1) & \cdots & F_n(\underline{x}_1) \\ F_1(\underline{x}_2) & F_2(\underline{x}_2) & \cdots & F_n(\underline{x}_2) \\ \vdots & \vdots & & \vdots \\ F_1(\underline{x}_n) & F_2(\underline{x}_n) & \cdots & F_n(\underline{x}_n) \end{bmatrix} \begin{bmatrix} a_1 \\ a_2 \\ \vdots \\ a_n \end{bmatrix} = \begin{bmatrix} f_1 \\ f_2 \\ \vdots \\ f_n \end{bmatrix}. \tag{5.2}$$

However, it is now possible to move any two nodes $\underline{x}_i$ and $\underline{x}_j$ so they end up with interchanged positions, without the nodes at any point having coincided. The effect on the matrix in (5.2) becomes that two rows have ended up interchanged, that is, the determinant has changed sign. By continuity, it must have been zero somewhere along the way.

Loosely speaking, no matter what fixed set of basis functions $F_k(\underline{x})$ are used (multivariate polynomials, trigonometric functions, combinations with exponentials, etc.), in more than 1-D, there are infinitely many configurations of distinct nodes for which interpolation becomes singular (and, with that, so do polynomial-based FD formulas). The classical remedy is to limit nodes to highly regular grid layouts,[3] as focused on in the previous Chapters 1–4. However, the present goal is complete geometric flexibility with regard to node layouts.

5.2 Moving Least Squares

5.2.1 Basic Idea for MLS

Referring to Figure 5.1, we consider the task of approximating the Laplacian $\Delta = \frac{\partial^2}{\partial x^2} + \frac{\partial^2}{\partial y^2}$ at the node location 1 by applying weights at nodes 1–8. In d spatial dimensions, there are

$$m = \binom{p+d}{d} \tag{5.3}$$

independent monomials up through degree p. With here $d = 2$, there are six independent monomials up through degree $p = 2$. For algebraic simplicity in finding weights, we can assume the stencil is translated such that $x_1 = y_1 = 0$ and take the monomials as $(1, x, y, x^2, xy, y^2)$. Finding coefficients $a_1, a_2, \ldots, a_6$ for these to match function values at the eight nodes leads to the overdetermined

[3] with each derivative approximation following a 1-D grid line.

linear system

$$\begin{bmatrix} 1 & x_1 & y_1 & x_1^2 & x_1y_1 & y_1^2 \\ 1 & x_2 & y_2 & x_2^2 & x_2y_2 & y_2^2 \\ 1 & x_3 & y_3 & x_3^2 & x_3y_3 & y_3^2 \\ \vdots & \vdots & \vdots & \vdots & \vdots & \vdots \\ \vdots & \vdots & \vdots & \vdots & \vdots & \vdots \\ \vdots & \vdots & \vdots & \vdots & \vdots & \vdots \\ \vdots & \vdots & \vdots & \vdots & \vdots & \vdots \\ 1 & x_8 & y_8 & x_8^2 & x_8y_8 & y_8^2 \end{bmatrix} \begin{bmatrix} a_1 \\ a_2 \\ \vdots \\ \vdots \\ \vdots \\ a_6 \end{bmatrix} = \begin{bmatrix} f_1 \\ f_2 \\ f_3 \\ \vdots \\ \vdots \\ \vdots \\ \vdots \\ f_8 \end{bmatrix}. \tag{5.4}$$

Standard methods for solving overdetermined systems in the least squares sense give the a-coefficients as linear combinations of the f-values. Applying the operator (here Δ) to these resulting polynomials gives FD weights.[4] In the present case of irregularly spaced nodes, this overdetermined system is highly unlikely to be ill conditioned (or singular).

It was recognized already when MLS approximations were first introduced by Lancaster and Salkauskas (1981) that the description above is oversimplistic, and the overdetermined linear system (5.4) needs to be enhanced by a weight function in order to produce FD approximations suited for PDE approximations.

5.2.2 Some Stencil Illustrations

Some key properties of MLS approximations can be illustrated already with the Laplacian Δ on unit-spaced grids in 2-D. The effect of applying Δ to a general 2-D Fourier mode $e^{i\omega_x x + i\omega_y y}$ is that it gets multiplied by the factor $-(\omega_x^2 + \omega_y^2)$, displayed in the left subplot in Figure 5.2(a). Approximating Δ with the standard 5-point and 9-point FD approximations

$$\Delta \approx \begin{bmatrix} & 1 & \\ 1 & -4 & 1 \\ & 1 & \end{bmatrix} / h^2 \quad \text{and} \quad \Delta \approx \begin{bmatrix} 1 & 4 & 1 \\ 4 & -20 & 4 \\ 1 & 4 & 1 \end{bmatrix} / (6h^2),$$

respectively, gives corresponding $[-\pi, \pi] \times [-\pi, \pi]$ – periodic surfaces in the ω_x, ω_y-plane as seen in the next two subplots along the top row in the figure. Below these are shown contour plots of their errors ($\log_{10}$ of the magnitude of the errors, compared to the analytic result in the left subplot in Figure 5.2(a)). While the error for the 5-point stencil is marginally smaller, the error is more isotropic for the 9-point stencil.

[4] As coefficients for the f-values.

5.2.3 Weight Functions for MLS

With Figure 5.2 as a reference, we consider next the 9-point stencil produced by the approach described in Section 5.2.1, which turns out to be

$$\Delta \approx \begin{bmatrix} 2 & -1 & 2 \\ -1 & -4 & -1 \\ 2 & -1 & 2 \end{bmatrix} /(3h^2),$$

displayed similarly in the left-hand column of subplots in Figure 5.3. This stencil has lost its "diagonal dominance,"[5] and high Fourier modes are grossly misrepresented. The remedy is to apply suitably weights to the equations in (5.4). For a square linear system, multiplying the equations by different (nonzero) constants will not affect the solution to the system. That is no longer the case for least squares solutions to overdetermined systems. For the present test problem, we can, for example, multiply equation k in the equivalent to (5.4) by $e^{-\varepsilon^2(x_k^2+y_k^2)}$, $k = 1, 2, \ldots, 9$, deemphasizing nodes further from the central one.[6] Using no multipliers (weight function) corresponds to $\varepsilon = 0$. The remaining three columns of subplots show much improved approximations, which for increasing ε first pass through a state of excellent isotropy before approaching the 5-point approximation described earlier.

This MLS approach applies immediately to scattered nodes and provides then weight sets that are well suited for scattered node FD-type stencils. The goal of this brief MLS summary was simply to convey its concept. The literature on MLS theory, algorithmic variations, and applications is nowadays very extensive. Much more thorough MLS discussions can be found for example in Fasshauer (2007), Chapters 22–27, and Wendland (2005), Chapter 4.

5.3 Global RBFs

The historical beginnings of radial basis functions (RBFs) and of subsequent RBF-generated FD (RBF–FD) methods are an evolution in the opposite direction to how local FD approximations led to global PS methods. RBFs started off as a spectrally accurate global interpolation method, featuring complete flexibility in the placing of its nodes. However, its computational cost grows unacceptably fast with the total number of nodes. From global RBFs then evolved the local RBF–FD approach, featuring FD-like ease of implementation and computational speeds on grid-free node layouts. Monographs that

[5] The central entry is in magnitude (much) less than the sum of the magnitudes of the other entries.

[6] Again for each stencil, we translate the central node to the origin.

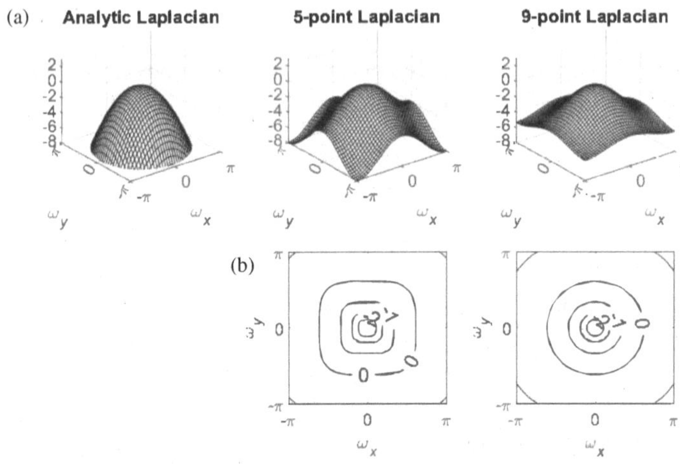

Figure 5.2 (a) The effect on Fourier modes $e^{i\omega_x x + i\omega_y y}$ by the analytic Laplacian and (b) contour plots of $\log_{10}$ of the errors (as described in Section 5.2.2).

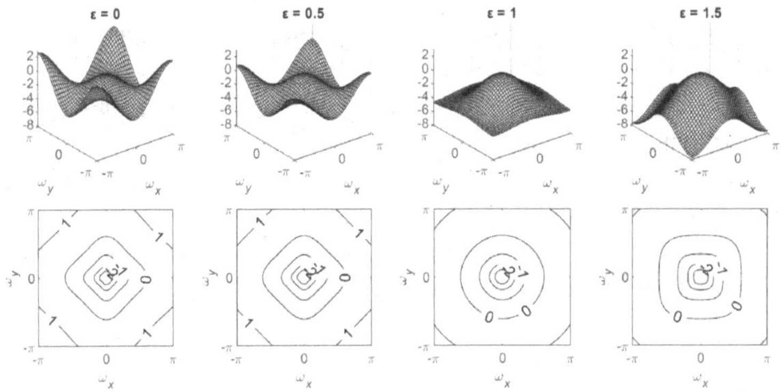

Figure 5.3 Displays similar to those in Figure 5.2, but here for MLS approximations with different weighting functions.

survey RBFs and their application to PDEs include Fasshauer (2007); Fornberg and Flyer (2015b), with the latter abbreviated in Fornberg and Flyer (2015c).

Although concepts similar to RBFs were used in statistics in the 1930s, their computational use for multivariate interpolation began with Rolland Hardy[7] and a cartography application in the early 1970s (Hardy, 1971).[8] It uses basis

[7] 1920–2009, American engineer and surveyor.

[8] A way to arrive at the RBF concept via traditional spline interpolation is described in Appendix B, Sections B.2.6–B.3.2.

Table 5.1 *Some common choices for radial functions $\phi(r)$.*

Type of basis function	**Radial function ϕ(r)**
Piecewise smooth RBFs	
Polyharmonic spline (PHS)	r^m, $m = 1, 3, 5, \ldots$
	$r^m \log r$, $m = 2, 4, 6, \ldots$
	$\frac{2^{1-m}}{\Gamma(m)} r^m K_m(\varepsilon r)$, $m > 0$ (K-Bessel function)
	Special cases: $e^{-\varepsilon r}$, $e^{-\varepsilon r}(1 + \varepsilon r)$ for $m = \frac{1}{2}, \frac{3}{2}$.
Compact support ("Wu–Wendland")	$(1 - \varepsilon r)_+^m p(\varepsilon r)$, p certain polynomials, $m = 1, 2, 3, \ldots$
Infinitely smooth RBFs	
Gaussian (GA)	$e^{-(\varepsilon r)^2}$
Multiquadric (MQ)	$\sqrt{1 + (\varepsilon r)^2}$
Inverse Multiquadric (IMQ)	$1/\sqrt{1 + (\varepsilon r)^2}$
Inverse Quadratic (IQ)	$1/(1 + (\varepsilon r)^2)$
Sech (SH)	$\operatorname{sech} \varepsilon r$
Bessel (BE) ($d = 1, 2, \ldots$)	$J_{d/2-1}(\varepsilon r)/(\varepsilon r)^{d/2-1}$

functions of the form $F_k(\underline{x}) = \phi(||\underline{x} - \underline{x}_k||)$, which are radially symmetric around center points $\underline{x}_k$, typically chosen to be the same as the node points at which interpolation data is provided. The only norm we will consider is the standard Euclidean 2-norm. Here $\phi(r)$ is a scalar function of a scalar variable. Many choices for $\phi(r)$ are in common use, with a few listed in Table 5.1.

In his pioneering work, Hardy used multiquadrics (MQ), and this remained for some time the most common choice. For all the "Infinitely smooth RBFs," ε is known as the shape parameter.[9]

5.3.1 RBF Concept Illustration

Choosing in (5.2) $F_k(\underline{x}) = \phi(||\underline{x} - \underline{x}_k||)$, the coefficients a_k in the RBF interpolant

$$s(\underline{x}) = \sum_{k=1}^{n} a_k \phi(||\underline{x} - \underline{x}_k||) \tag{5.5}$$

[9] Much literature has been devoted to problem-specific optimization of ε. However, we can here bypass that subject, as the RBF–FD methods we will focus on mainly use PHS of the form $\phi(r) = r^m$, with m odd, which are shape parameter free (if a shape parameter ε was applied, this would scale the expansion coefficients in such a way that its effect would get canceled out).

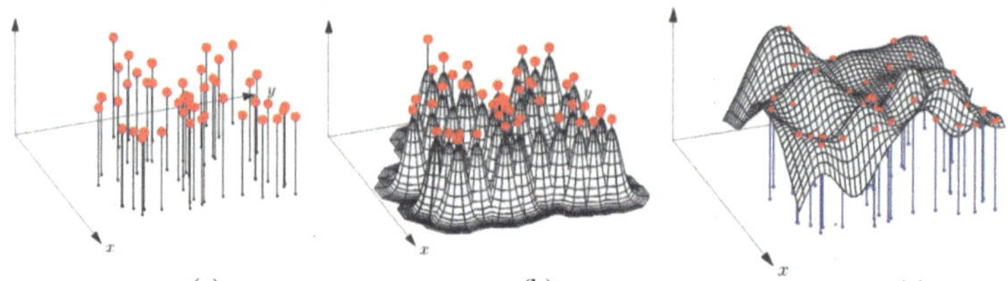

(a) (b) (c)

Figure 5.4 Graphical illustration of the three steps in the RBF process of interpolating scattered data. Figure adapted from Fornberg and Flyer (2015c).

to the data $f(x_k) = f_k$, $k = 1, 2, \ldots, n$, (cf. (5.2)) will be obtained by solving

$$\begin{bmatrix} \phi(\|\underline{x}_1 - \underline{x}_1\|) & \phi(\|\underline{x}_1 - \underline{x}_2\|) & \cdots & \phi(\|\underline{x}_1 - \underline{x}_n\|) \\ \phi(\|\underline{x}_2 - \underline{x}_1\|) & \phi(\|\underline{x}_2 - \underline{x}_2\|) & \cdots & \phi(\|\underline{x}_2 - \underline{x}_n\|) \\ \vdots & \vdots & & \vdots \\ \phi(\|\underline{x}_n - \underline{x}_1\|) & \phi(\|\underline{x}_n - \underline{x}_2\|) & \cdots & \phi(\|\underline{x}_n - \underline{x}_n\|) \end{bmatrix} \begin{bmatrix} a_1 \\ a_2 \\ \vdots \\ a_n \end{bmatrix} = \begin{bmatrix} f_1 \\ f_2 \\ \vdots \\ f_n \end{bmatrix}. \tag{5.6}$$

The idea behind this is illustrated in Figure 5.4. Part (a) illustrates function values at a set of scattered node points in the x, y-plane. In Part (b), a Gaussian RBF (shown quite peaked, i.e., with a fairly large value of the shape parameter ε) has been centered at each of these. Part (c) shows the result when all these Gaussians have been added using the coefficients a_k obtained from solving (5.6), which ensures that the linear combination obeys all the given data. Since each of the Gaussians is infinitely differentiable, so becomes the resulting interpolant $s(\underline{x})$. The use of analytic derivatives of this type of interpolants for spatial differentiation of PDEs was proposed by Kansa (1990a,b).[10]

5.3.2 Some Nonsingularity Results for RBF Interpolation

It transpires that, for many choices of radial function $\phi(r)$, the system (5.6) (in contrast to (5.2)) can never be singular, no matter how any number of distinct nodes $\underline{x}_k$ are scattered in any number of dimensions. Since the matrix in (5.6) will appear several times in what follows, it will often be referred to as the RBF matrix, or shorter still, the A-matrix.

The singularity demonstration above for multivariate polynomials (Section 5.1) no longer applies, since exchanging the places of two nodes $\underline{x}_i$ and $\underline{x}_j$ for the A-matrix exchanges *both two rows and two columns*, leaving the sign of the determinant unchanged. This observation alone is of course insufficient to

[10] focusing on MQ-type RBFs. The state of the art for RBF approximation theory at this time is summarized in Powell (1992).

show unconditional nonsingularity for nodes that are distinct. Strict proofs for that go back to the 1930s, when matrices of the same structure arose in other contexts (Bochner, 1933; Schoenberg, 1938), and are repeated in all main RBF monographs. Most proofs start by considering the case of the GA radial function $\phi(r) = e^{-(\varepsilon r)^2}$ (e.g., Fornberg and Flyer, 2015b, Section 3.1.3 and Fasshauer, 2007, Section 3.2). After rewriting each entry in the A-matrix as the inverse of its Fourier transform, it can, in relatively few steps, be established that A becomes a positive-definite matrix. Using some additional mathematical tools (Laplace transforms and some theory on 'completely monotone functions'), nonsingularity then follows also for several other RBF types. These include all the infinitely smooth radial functions listed in Table 5.1.[11]

5.3.3 Some Comments on the RBF Shape Parameter ε

Many of the RBF types have a shape parameter ε available for problem-dependent tuning, and the choice of it will greatly affect both resulting accuracy and computational stability. For simplicity, we focus in this section on GA RBFs.[12]

5.3.3.1 Accuracy as Function of Shape Parameter ε

The character of the solid curve in Figure 5.5 was noted early in the development of RBFs (Tarwater, 1985). Use of flatter RBFs (smaller ε) increases the accuracy until a breakdown occurs due to ill conditioning of the A-matrix in (5.6).[13]

The shape of this solid curve raises some obvious questions:

i. What error levels could be reached if it was not for the numerical ill conditioning?
ii. Can the accuracy breakdown seen in the jagged part of the solid curve in Figure 5.5(b) be avoided? In particular, is it possible to reach full double precision (DP; $O(10^{-16})$) accuracy if doing all calculations in DP (i.e., not resorting to extended precision arithmetic)?

[11] Some fine print: (i) The proof of nonsingularity is somewhat different in the MQ case, as their A-matrices have one positive eigenvalue with all the remaining ones negative (Micchelli, 1986; Powell, 2005). (ii) In the case of BE, nonsingularity is only assured in up to d spatial dimensions ($d \geq 2$).

[12] For some of the other RBF types, minor additional fine print may be needed; see Fornberg and Flyer (2015b), Section 3.2.

[13] Much effort in the earlier RBF literature was devoted to approximating the transition point in the solid curve, that is, finding an "optimal" (application-dependent) value for the shape parameter ε. However, we describe below some ways that make this issue mostly obsolete.

Regarding (i), extrapolation of the smooth part of the error curve toward lower ε might suggest that the accuracy for $\varepsilon \to 0$ should improve indefinitely. That is, however, impossible as only a finite amount of data is provided. For very small ε, there arises another limiting factor, explained in terms of the polynomial Runge phenomenon in Fornberg and Flyer (2015b), Section 3.3.

Regarding (ii), there remained for some time significant confusion, as an "uncertainty principle" was formulated, asserting that the ill conditioning of the A-matrix was fundamental to the concept of RBFs, requiring an inevitable compromise between conditioning and accuracy (Schaback, 1995). It was, however, discovered (Driscoll and Fornberg, 2002) that, in 1-D, and for most smooth RBF types, the RBF interpolant in the flat $\varepsilon \to 0$ limit simply becomes the Lagrange interpolating polynomial (with counterpart limits for higher-D).[14] Important conclusions from this result are:

i. RBF interpolation in the $\varepsilon \to 0$ limit on a Cartesian grid will typically reproduce PS methods[15].
ii. The RBF-Direct approach (solving (5.6) followed by evaluating (5.5)) is merely an ill-conditioned numerical algorithm for a genuinely well-conditioned problem.
iii. There ought to exist numerical algorithms that remain well conditioned even as $\varepsilon \to 0$.

Regarding (i), one can note that RBF interpolants are not restricted to grids, and also (as seen in Figure 5.5(b)), the $\varepsilon \to 0$ limit need not be optimal for accuracy. Attempting to overcome (ii) by using extended precision arithmetic is unlikely to be cost-effective if it requires changing from hardware- to software-provided arithmetic.[16] Section 5.3.3.2 below follows up on observation (iii). It should also already now be noted that, when soon turning to RBF–FD, the main discretization method (PHS+poly, being shape parameter-free) does not have this type of conditioning issues.

5.3.3.2 Stable Algorithms

Stable algorithms that are practical for numerical computations (and not primarily for theoretical insights) fall in two categories:

[14] In this flat limit, the coefficients a_k, $k = 1, 2, \ldots, n$, will diverge toward $\pm\infty$. However, off-setting cancellations will occur in (5.5). While both (5.6) and (5.5) become ill-conditioned computational steps, the end result is not.

[15] Some fine print is needed for a more definite statement.

[16] The level of arithmetic precision needed increases rapidly as $\varepsilon \to 0$, as analyzed in Fornberg and Flyer (2015b), Section 3.2.1. Actually reaching $\varepsilon = 0$ (which, if considering ε complex valued, can be seen as a well-posed case of a removable singularity), is unachievable with RBF-Direct (solving (5.6) followed by evaluating (5.5)) and extended precision, no matter how many digits are used.

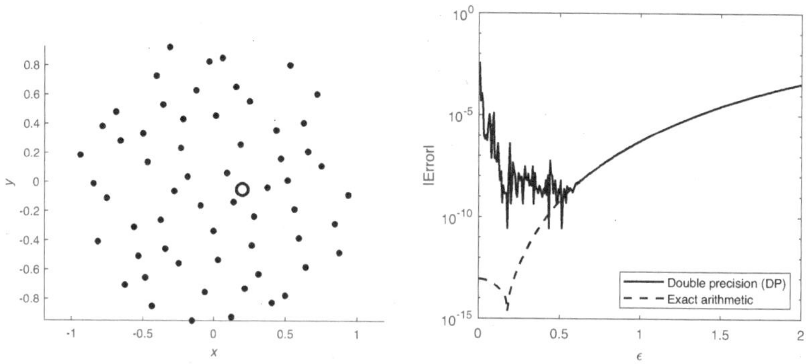

(a) Data points/RBF centers and an evaluation point

(b) Error at evaluation point as a function of ε

Figure 5.5 RBF interpolation error as function of ε, using GA RBFs, for the test problem shown in Part (a): Interpolate $f(x, y) = 59/(67 + (x + \frac{1}{7})^2 + (y - \frac{1}{11})^2)$ from the displayed points to obtain an approximation at the circular marker. Part (b): Solid curve: Using standard double precision (DP), Dashed curve: Using exact arithmetic, approximated by either extended precision or a stable algorithm. In this example, DP is seen to fail around $\varepsilon = 0.5$, whereas quad precision (128 bit floating point) will fail around $\varepsilon = 0.2$.

i. Utilize the fact that $s(\underline{x}, \varepsilon)$ (as defined by (5.5), (5.6)) is an analytic function of ε, if the radial function $\phi(r, \varepsilon)$ is analytic in ε (complex). The point $\varepsilon = 0$ then typically becomes a removable singularity of $s(\underline{x}, \varepsilon)$. The historically first stable algorithm, RBF–CP, standing for RBF–Contour–Padé, (Fornberg and Wright, 2004) utilized this concept. It is improved on by the more recent RBF–RA (RBF–Rational Approximation), algorithm (Wright and Fornberg, 2017).

ii. Change basis within the same space as spanned by a fixed set of near-flat RBFs. Algorithms in this category go by the abbreviations RBF–QR (Fornberg et al., 2011; Fornberg and Piret, 2007), and RBF–GA (Fornberg et al., 2013).[17] The general concept is illustrated schematically in Figure 5.6. Similarly to these two illustrations (of vectors in 3-D space, and polynomials over $[-1, 1]$), the function space spanned by near-flat RBFs can also be spanned by well-conditioned basis functions. To find these alternate basis

[17] RBF–GA is currently the computationally fastest available stable algorithm, with computer times about 10 times those for RBF–Direct. Expansions based on Mercer's theorem, as considered in Fasshauer and McCourt (2012), are more focused on theoretical insights than on computational performance. Other attempts to stabilize GA interpolation for small ε-values include the use of basis modifications with radial polynomials (Mishra et al., 2018; Pooladi and Larsson, 2024). A main challenge is to obtain both high accuracy and assured nonsingularity.

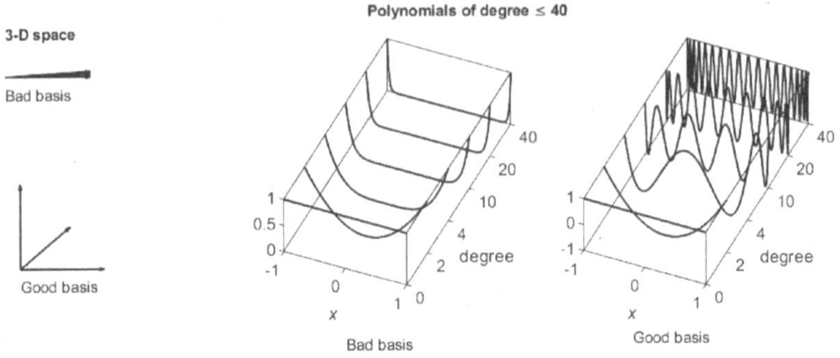

Figure 5.6 Two examples of bad (ill-conditioned) and good (well-conditioned) basis sets that span the same space. The polynomial bases illustrated are monomials x^n and Chebyshev polynomials $T_n(x)$.

functions, the key steps must be performed analytically (rather than by immediate numerics).

5.3.4 Differentiation Matrices

The process of approximating a partial derivative operator L based on scattered data can be achieved by first obtaining RBF expansion coefficients by solving (5.6), and then summing the equivalent of (5.5) in which each basis function $\phi(\|\underline{x} - \underline{x}_k\|)$ has been replaced by $L\,\phi(\|\underline{x} - \underline{x}_k\|)$. Especially if this linear process needs to be repeated large number of times (such as during a time stepping procedure), it is much more efficient to express it in the form of a differentiation matrix (DM):

$$L \begin{bmatrix} u_1 \\ u_2 \\ \vdots \\ \vdots \\ u_N \end{bmatrix} = \begin{bmatrix} & & \\ & & \\ & \text{DM} & \\ & & \\ & & \end{bmatrix} \begin{bmatrix} u_1 \\ u_2 \\ \vdots \\ \vdots \\ u_N \end{bmatrix}. \tag{5.7}$$

When using global RBFs, both the RBF matrix in (5.6) and the DM in (5.7) are full matrices. Using compactly supported RBFs (e.g., Wu–Wendland functions, briefly noted in Table 5.1) gives sparsity to the RBF matrix, but the resulting DMs are still full matrices. The main benefit of RBF–FD approximations (introduced in Section 5.4) is that these lead to highly sparse DM matrices.

5.3.5 Strengths and Limitations of Global RBFs

Only in rare cases can the benefits of spectral accuracy offset the high computational cost of using global RBFs (especially in higher dimensions). Nevertheless, the approach produced, shortly before the more widespread use of localized RBF–FD discretizations, the arguably first instance where an RBF calculation was not just applied to a "toy problem" about which everything was known already, but shown to compare very favorably against all previous approaches on an extensively benchmarked large-scale problem.

A calculation of mantle flow in the earth: The RBF calculation reported in Wright et al. (2010) concerned fine-structured viscous flows in the earth's mantle, driven by the temperature difference between the outer core and the crust.[18] Figure 5.7 shows two snapshots of a movie,[19] with yellow corresponding to regions with temperatures a certain amount higher than their surroundings (upwellings), and blue to similar downwellings. Extensive comparisons were presented, showing the RBF method to be the most effective available. It additionally produced some unexpected novel physical insights. The spherical shell region was discretized using ME (minimal energy) nodes distributed over a number of spherical shells, using global MQ RBFs over each shell together with Chebyshev discretizations radially (giving spectral accuracy in all directions). A key advantage of using scattered nodes over spherical shells (represented by their x, y, z-coordinates) is that these encounter no "pole problems," which arise whenever any surface-bound coordinate system is used (e.g., the north and south pole coordinate singularities in spherical coordinates).[20]

5.3.6 RBFs with Supplementary Polynomials

A commonly used variant of (5.5) is

$$s(\underline{x}) = \sum_{k=1}^{n} a_k \phi(||\underline{x} - \underline{x}_k||) + \sum_{k=1}^{m} \gamma_k p_k(\underline{x}), \tag{5.8}$$

where the $p_k(\underline{x})$ form a basis for all polynomials up through some degree p in the d components of the $\underline{x}$ vector (i.e., when approximating in a d-dimensional space). The number of these polynomial terms is $m = \binom{p+d}{p}$. With now m additional coefficients γ_k, one obtains again a square linear system to solve by

[18] Modeled by a Boussinesq fluid at infinite Prandtl number.

[19] Computed by the first author of Wright et al. (2010).

[20] Even if the governing equations are formulated in spherical coordinates, pole singularities disappear when the PDEs are transformed to the form of RBF DMs.

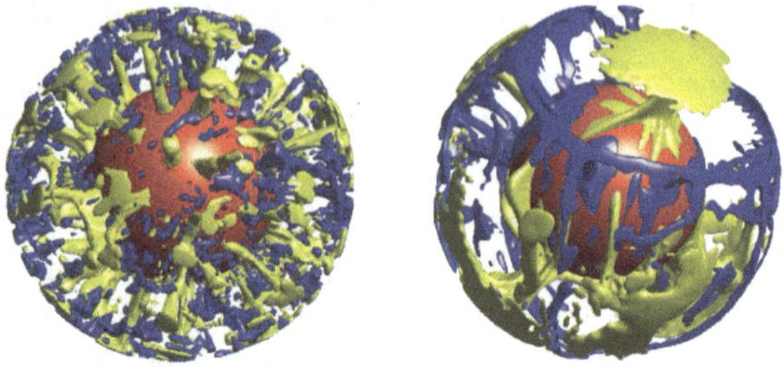

Figure 5.7 Two snapshots from a mantle convection movie, calculated as described in Wright et al. (2010).

imposing m matching constraints on the RBF coefficients a_k:

$$\sum_{k=1}^{n} a_k p_r(\underline{x}_k) = 0, \quad r = 1, 2, \ldots, m. \tag{5.9}$$

Example 5.3.1 Write out explicitly the linear system described by (5.8), (5.9) in the case of $p = 2$ (i.e., including up to quadratic terms) in $d = 2$ space dimensions.

Solution For $p = 2$ and $d = 2$, $m = \binom{4}{2} = 6$ terms are added and equally many constraints are imposed. Writing $\underline{x} = \left[\begin{smallmatrix} x \\ y \end{smallmatrix}\right]$, (5.8) becomes

$$s(\underline{x}) = \sum_{k=1}^{n} a_k \phi(||\underline{x} - \underline{x}_k||) + \gamma_1 + (\gamma_2 x + \gamma_3 y) + (\gamma_4 x^2 + \gamma_5 xy + \gamma_6 y^2) \tag{5.10}$$

subject to the six constraints

$$\sum_{k=1}^{n} a_k 1 = \sum_{k=1}^{n} a_k x_k = \sum_{k=1}^{n} a_k y_k = \sum_{k=1}^{n} a_k x_k^2 = \sum_{k=1}^{n} a_k x_k y_k = \sum_{k=1}^{n} a_k y_k^2 = 0. \tag{5.11}$$

With function values f_k, $k = 1, 2, \ldots, n$, the linear system that needs to be solved to determine the coefficients in (5.10) becomes

$$
\left[\begin{array}{ccc|cccccc}
 & & & 1 & x_1 & y_1 & x_1^2 & x_1 y_1 & y_1^2 \\
 & A & & \vdots & \vdots & \vdots & \vdots & \vdots & \vdots \\
 & & & \vdots & \vdots & \vdots & \vdots & \vdots & \vdots \\
 & & & 1 & x_n & y_n & x_n^2 & x_n y_n & y_n^2 \\
\hline
1 & \cdots & 1 & & & & & & \\
x_1 & \cdots & x_n & & & & & & \\
y_1 & \cdots & y_n & & & & 0 & & \\
x_1^2 & \cdots & x_n^2 & & & & & & \\
x_1 y_1 & \cdots & x_n y_n & & & & & & \\
y_1^2 & \cdots & y_n^2 & & & & & &
\end{array}\right]
\left[\begin{array}{c} a_1 \\ \vdots \\ \vdots \\ a_n \\ \hline \gamma_1 \\ \gamma_2 \\ \gamma_3 \\ \gamma_4 \\ \gamma_5 \\ \gamma_6 \end{array}\right]
=
\left[\begin{array}{c} f_1 \\ \vdots \\ \vdots \\ f_n \\ \hline 0 \\ 0 \\ 0 \\ 0 \\ 0 \\ 0 \end{array}\right],
\tag{5.12}
$$

where A is the same matrix as in (5.6).[21] In block form, we write (5.12) as

$$
\begin{bmatrix} A & P^T \\ P & 0 \end{bmatrix} \begin{bmatrix} \underline{a} \\ \underline{\gamma} \end{bmatrix} = \begin{bmatrix} \underline{f} \\ \underline{0} \end{bmatrix}. \tag{5.13}
$$

□

This RBF formulation with supplementary polynomials comes with a range of advantages – many of which become especially important in the following context of using RBFs locally (in RBF–FD stencils):

i. In cases of A positive definite, the system (5.13) is the same as one would obtain using Lagrange multipliers (cf. Appendix D) to minimize the quadratic form $\frac{1}{2}\underline{a}^T A \underline{a} - \underline{a}^T \underline{f}$ subject to the constraint equations $P\underline{a} = \underline{0}$. This ensures that the RBF coefficients in $\underline{a}$ stay small (and that the system remains nonsingular) (Bayona, 2019b; Bayona et al., 2019).
ii. In the case of using PHS-type radial functions (cf. Table 5.1), for which the A-matrix by itself might be singular, the system (5.13) is nonsingular if the P-matrix is of full rank.[22]
iii. The added polynomials greatly reduce Runge phenomenon-type boundary effects (as described next in Section 5.3.7).

[21] The graphical layout in (5.12) is slightly misleading since, for nonsingularity, the square A-block must be larger in size than the square 0-block.

[22] This typically becomes the case if the nodes are not lattice based, and N exceeds the number m of monomial terms (i.e., P has more columns than rows).

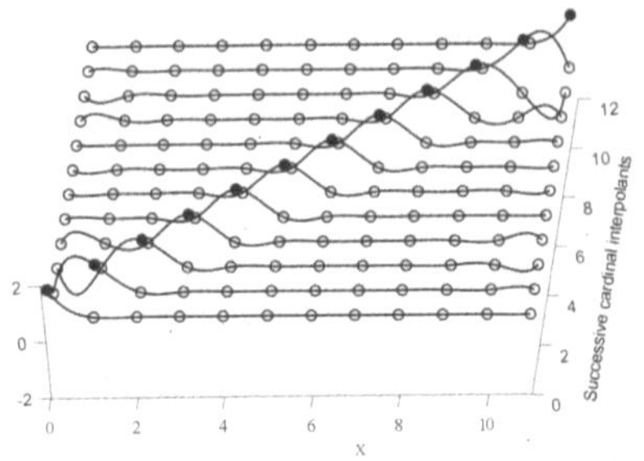

(a) Set of 11 cardinal functions

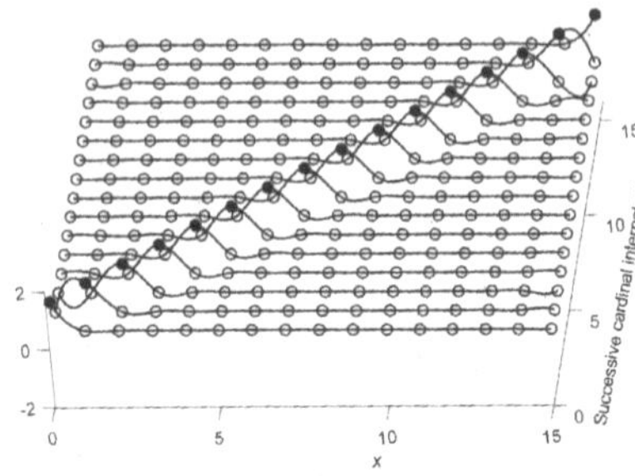

(b) Set of 15 cardinal functions

Figure 5.8 Cardinal functions when using $\phi(r) = |r|^3$ together with supplementary polynomials up through degree 8.

5.3.7 Cardinal Functions

Figure 1.3 showed degree-8 polynomial interpolants to cardinal data. The large oscillations in several of these lead to the Runge phenomenon (RP), seen in Figure 1.4, severely limiting the utility of polynomial interpolants on equispaced data. Using RBFs (such as cubics $\phi(r) = |r|^3$) together with the same set of polynomials dramatically reduces the RP in any number of space dimensions. For graphical simplicity, we illustrate this again in 1-D, in Figure 5.8.

These functions will again interpolate exactly all polynomials up through degree 8, in which case all the RBFs will have vanished. When using a PHS+poly-based RBF–FD stencil increasingly locally on a smooth function, the convergence rate will be entirely determined by the polynomial degree, and the fact that $\phi(r) = |r|^3$ has a discontinuous third derivative becomes irrelevant to the convergence order. A heuristic reason for the excellent cardinal functions seen in Figure 5.8 is that combinations of the $\phi(r) = |r|^3$ amount to cubic splines, and are thus well able to handle the cardinal aspect of the data (see the last subplot in Figure B.4). The similarity between (5.12) and (D.12) will ensure that the RBF coefficients otherwise are minimized.[23]

5.4 RBF-Generated Finite Differences

As for regular FD, we want for each node to weigh together function values at nearby nodes, to obtain derivative approximations. However, the nodes are no longer grid based, and the stencils generally all differ from each other in the

[23] For more thorough discussions of PHS+poly cardinal functions, see, for example, Bayona et al. (2019); Bayona (2019b).

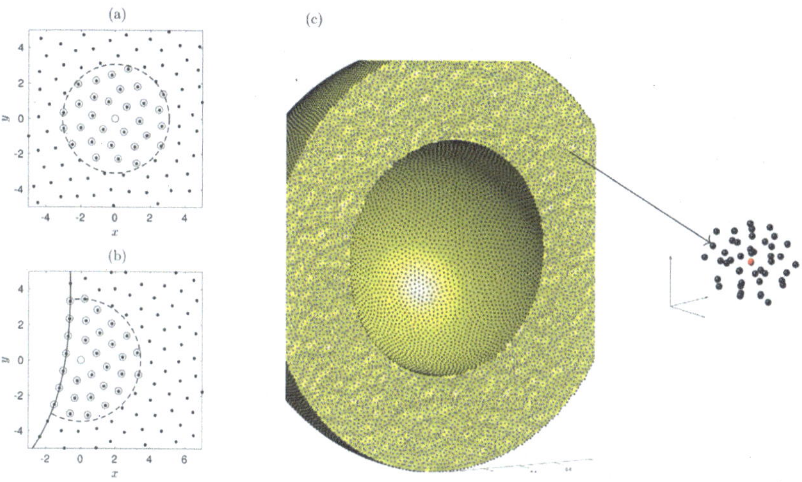

Figure 5.9 Illustrations of RBF–FD stencils: (a), (b) Stencils in 2-D domain interior and near to a boundary, respectively (the observations in Section 5.3.7 greatly reduce any need for node refinement at the boundary); (c) 3-D stencil within a spherical shell. Parts (a), (b) are reproduced from Bayona et al. (2019) and Part (c) from Wright and Fornberg (2017).

details of their node layouts. Figure 5.9 gives some 2-D and 3-D illustrations. Each stencil requires its own set of weights – fortunately a trivially parallelizable task.

Traditional FD weights, as described in Section 1.2, are determined to give the exact result for monomials of as high degree as possible. When moving from a grid in 1-D to scattered nodes in higher dimensions, two additional issues arise if using polynomials only:

i. As noted earlier, there are $\binom{p+d}{p}$ polynomials of degree p in d-D. If the number of nodes in a stencil does not match one of these numbers, which polynomial terms should be included versus omitted?
ii. Even if the counts match, the calculation of weights will become singular (i.e., weights diverge to infinity) for many node configurations (cf. Section 5.1).

Both these issues are bypassed if one uses RBFs in place of polynomials in the weights calculations. This was first suggested in a conference presentation (Tolstykh, 2000). Independently, several other researchers (unaware of that presentation) came up with the same idea in the next few years, leading to what is now known as the RBF–FD method.[24] This generalizes standard FD methods in the sense that, if one uses, for example, Gaussian RBFs and lets $\varepsilon \to 0$ (making them flatter), the resulting interpolants approach a certain minimal degree polynomial form. For nodes on a grid, standard FD formulas are typically recovered. However,

i. small ε-values give ill-conditioned linear systems that may need the use of a stable algorithm, and
ii. pushing ε all the way to zero may not be optimal for accuracy (see the dashed curve in Figure 5.5(b)).

The shape parameter-free RBF–FD PHS+poly approach, described next, bypasses these issues.

5.4.1 RBF–FD Using PHS+Poly

After having created a suitable quasi-uniform node set (cf. Appendix G), one decides on a stencil size n. The kd-tree algorithm (in MATLAB implemented in the knnsearch function in the Statistics and Machine Learning Toolbox) provides the n nearest neighbors to each of the N nodes in roughly $O(Nn \log N)$ operations. Separately for each stencil, we then consider an interpolant of the form (5.10), (5.11), with a polyharmonic spline radial function, typically $\phi(r) = r^3, r^5$, or r^7, together with $m = \binom{p+d}{p}$ polynomials up through degree p, and with the matching constraints. Suppose the nodes $\underline{x}_1, \underline{x}_2, \ldots, \underline{x}_n$ in the stencil are ordered such that $\underline{x}_1$ is the stencil's *center node* (the one at which we want to approximate an operator L (such as $\frac{\partial}{\partial x}$, $\frac{\partial^2}{\partial x^2} + \frac{\partial^2}{\partial y^2}$, etc.), the weights are obtained by solving the linear system

$$\begin{bmatrix} A & P^T \\ P & 0 \end{bmatrix} \begin{bmatrix} \underline{w} \\ \underline{\lambda} \end{bmatrix} = \begin{bmatrix} \underline{L} \\ \underline{p} \end{bmatrix}, \tag{5.14}$$

where the coefficient matrix $\left[\begin{smallmatrix} A & P^T \\ P & 0 \end{smallmatrix}\right]$ is exactly as in (5.12), (5.13), $\underline{L}$ is a column vector with the entries $L(\phi(||\underline{x} - \underline{x}_k||))\big|_{\underline{x}=\underline{x}_1}$ for $k = 1, 2, \ldots, n$, and $\underline{p}$ is a column vector with $L(p_r(\underline{x}))\big|_{\underline{x}=\underline{x}_1}$ for $r = 1, 2, \ldots, m$. Solving the system gives $\underline{w}$ (a column vector of the n desired RBF–FD weights) and $\underline{\lambda}$ (a vector of m coefficients, which typically is discarded, can be interpreted as Lagrange multipliers). Since the coefficient matrix is the same as for RBF interpolation,

[24] Some literature refers to RBF–FD as RBF–DQ (with DQ standing for differential quadrature) or as the radial point method.

Algorithm 5.1 MATLAB code for the 2-D case of finding PHS+poly stencil weights for $\frac{\partial}{\partial x}$, $\frac{\partial}{\partial y}$, and $\frac{\partial^2}{\partial x^2} + \frac{\partial^2}{\partial y^2}$.

```
function w = RBF_FD_PHS_pol_weights (x,y,m,d)

% Input parameters
%   x,y     Column vectors with stencil node locations; approximation
%           to be accurate at x(1),y(1)
%     m     Power of r in RBF fi(r) = r^m, with m odd, >= 3.
%     d     Degree of supplementary polynomials (d = -1 no polynomials)
%
% Output parameter
%     w     Matrix with three columns, containing RBF-FD weights for
%           d/dx, d/dy, and the Laplacian d2/dx2+d2/dy2, respectively.
x = x-x(1); y = y-y(1);    % Shift nodes so stencil centered at origin
n = length(x);

% -- RBF part ------------------
A0 = sqrt((x-x').^2+(y-y').^2).^m;    % RBF A-matrix
L0 = m * (sqrt(x.^2+y.^2).^(m-2)).*[-x,-y,m*ones(n,1)] ;% RHSs

% -- Polynomial part ---------------
if d == -1                        % Special case; no polynomial terms
     A = A0; L = L0;              % i.e. pure RBF
else   % Create matrix with polynomial terms and matching constraints
     X = x(:,ones(1,d+1)); X(:,1) = 1; X = cumprod( X,2);
     Y = y(:,ones(1,d+1)); Y(:,1) = 1; Y = cumprod( Y,2);
     np = (d+1)*(d+2)/2;            % Number of polynomial terms
     XY = zeros(n,np); col = 1;    % Assemble polynomial matrix block
     for k = 0:d
         XY(:,col:col+k) = X(:,k+1:-1:1).*Y(:,1:k+1);
         col = col+k+1;
     end
     L1 = zeros(np,3);  % Create matching RHSs
     if d >= 1; L1(2,1) = 1; L1(3,2) = 1; end
     if d >= 2; L1(4,3) = 2; L1(6,3) = 2; end
     A = [A0,XY;XY',zeros(col-1)];  % Assemble lin. system to be solved
     L = [L0;L1];                   % Assemble RHSs
end

% -- Solve for weights ---------------
W = A\L;
w = W(1:n,:);
end
```

the nonsingularity argument from above again applies, and it is again ensured that the norm of the solution vector (now $\underline{w}$) stays small.

The MATLAB code listed as Algorithm 5.1 is for the 2-D case of finding the weights for $\frac{\partial}{\partial x}$, $\frac{\partial}{\partial y}$, $\frac{\partial^2}{\partial x^2} + \frac{\partial^2}{\partial y^2}$, in a single stencil (where the first node is taken as the center node – just as node pointers are provided by MATLAB's kd-tree algorithm `knnsearch`).

The following is a concise list of observations concerning the RBF–FD PHS+poly discretization approach (Flyer et al., 2016)[25]:

i. The formal order of accuracy/convergence rate will be determined by the polynomial degrees only (i.e., one easily realizes quite high orders, such as 4–8).

ii. The role of the supplementary polynomials is very different between global RBFs and RBF–FD. In the former case, there is only one polynomial. While this suffices for establishing nonsingularity, it is insufficient to contribute much to the overall accuracy. In the RBF–FD case, there is a separate polynomial for each stencil. It is now these that determine the order of accuracy (under refinement), with the RBFs then "fading out" (although still ensuring nonsingularity).

iii. A heuristic reason for the RBFs "fading out" under node refinement is as follows: The polynomials then get increasingly able to alone represent a smooth target function. If they were able to do this perfectly, we would have $P^T \underline{\gamma} = \underline{f}$ in (5.13) and, since (5.13) has a unique solution, $\underline{\alpha} = \underline{0}$, that is, the RBFs would be gone entirely. Two further observations related to this: (i) using, say $\phi(r) = r^3$, which is only twice differentiable, will not damage a higher order of accuracy that might be provided by the supplementary polynomials; and (ii) PHS+poly approximations are attractive at domain boundaries (as discussed in Section F.3).

iv. The PHS+poly approach is particularly easy to use, as it is shape parameter-free (as we already have noted repeatedly). The condition number of the matrix in (5.14) may become very large, but this is more a shortcoming in the definition of the condition number than any problem in obtaining accurate solutions using a standard linear system solver (which starts by rescaling the equations, such as "\" in MATLAB).[26]

v. The cardinal functions in Figure 5.8 are free from the spurious oscillations seen near the interval ends in Figure 1.3. This translates into much better one-sided RBF–FD stencils than in the case of regular FD approximations, as discussed further in Section F.3.

[25] It has additionally been reported in Santos et al. (2018) that for RBF–FD solutions of elliptic equations on irregular node sets, PHS+poly has for matching accuracy an edge over RBF-GA (cf. Section 5.3.3.2) with small ε in terms of computational speed, numerical stability, and code simplicity.

[26] This issue of systems of this structure featuring misleading condition numbers was first observed in Iske (2003).

5.4.2 RBF–FD Differentiation Matrices

When using grid-based FD methods, it is often unnecessary to explicitly form *differentiation matrices* (DMs). However, in the context of RBF–FD approximations, DMs are very convenient, as they also contain the information about which nodes are included in each stencil. In RBF–FD DMs (cf. (5.7)), each of the N rows will have n nonzero entries, that is, with $N \gg n$ this is a highly sparse matrix (and should be stored in a sparse format). If MATLAB's `knnsearch` has been used to identify the nearest neighbors to each node, it will also have provided indices for the positions in each row where the n RBF–FD weights are to be placed.

Before proceeding to use such a DM numerically, it is usually very beneficial to reorder (reindex) the nodes, so that the nontrivial DM entries become clustered around its main diagonal. A variety of strategies are available for this. Figure 5.10(a) shows a typical sparsity pattern in the case of $N = 6,400$ MD (maximal determinant) nodes on the surface of a sphere,[27] and stencil size $n = 50$ (i.e., there are here precisely 50 nonzero entries on each of the DM's 6,400 rows). After reordering the nodes using the Reverse Cuthill–McKee algorithm (in MATLAB `symrcm`), this pattern becomes as seen in Part (b). If one wishes to solve a linear system with the DM as coefficient matrix using a direct solver, the fill-in has now become vastly reduced (as this will stay within the envelope of the DM entries). When using the DM for operations such as matrix $\times$ vector multiplications,[28] a sparsity pattern as in Part (c) may be even more advantageous (due to how data in computer memories is typically accessed). The matrix structure shown here was obtained using a "locality sensitive hashing algorithm" (Bollig et al., 2012).

5.4.3 Stabilization Using Hyperviscosity

The main idea can be illustrated already on equispaced grids in 1-D, as shown in Figure 5.11. The dotted curves correspond to the curves in Figure 2.1 but are labeled with stencil width n rather than accuracy order $p = n - 1$. As noted before, only a narrow range of modes around $\omega = 0$ are treated accurately. Modes that deviate from the black diagonal line should preferably be eliminated. Viscous damping can achieve this. As the stencil width n is increased from 3 to $5, 7, 9, \ldots$, the diffusive term can be chosen to approximate $+\partial^2/\partial x^2$, $-\partial^4/\partial x^4$, $+\partial^6/\partial x^6$, $-\partial^8/\partial x^8, \ldots$. resulting in the sequence of dashed curves in the figure (causing an amount of damping proportional to the curve's elevation above the horizontal axis; all shown in the figure normalized to take the value 1 at

[27] obtained from the tables in Womersley (2020).
[28] for example when using iterative linear system solvers, or carrying out explicit time stepping.

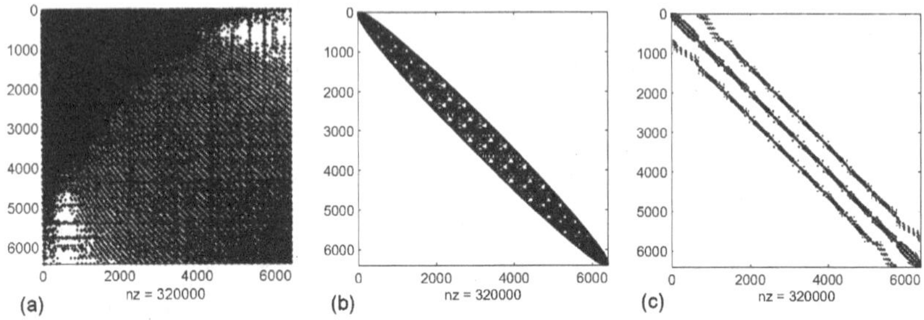

Figure 5.10 Comparisons of RBF–FD-produced DM sparsity patterns (reproduced from Fornberg and Flyer (2015b)).

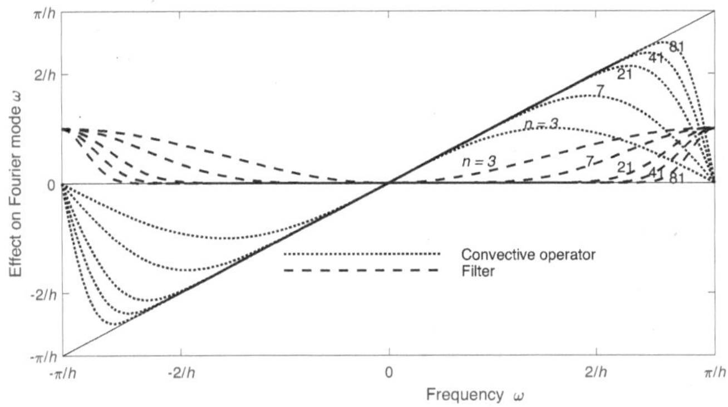

Figure 5.11 Spectral properties of convective versus diffusive operators on a 1-D grid with spacing h and with n denoting the stencil width.

$\omega = \pm\pi/h$). The diffusion term needs to be multiplied by some factor γ to obtain a suitable compromise between adequately removing (likely unstable) high-frequency modes and not causing instabilities by increasing the stiffness of the time stepping too much. The range of accurately treated Fourier modes is seen to increase with n roughly at the same rate as the range of modes increases that hyperviscosity leaves intact. It thus makes sense to use the same width stencils for both the convective term(s) and for the hyperviscosity. The concept just illustrated carries over quite directly over to RBF–FD stencils on quasi-uniform node sets (Fornberg and Lehto, 2011), then with the diffusion approximating powers of the Laplacian instead of even order unidirectional derivatives. The main practical issue is to select the γ-parameter. Guidelines for this are discussed in Flyer et al. (2012); Shankar and Fogelson (2018).

Reproducing one of the RBF–FD hyperviscosity demonstration test cases in Fornberg and Lehto (2011), Figure 5.12 shows on an "unrolled" sphere (i.e., over a longitude φ, latitude θ plane) the end result, interpolated to a Cartesian grid, after time stepping (with RK4) a cosine bell through 1,000 full transits around the sphere. The node set consisted of $N = 25,600$ quasi-uniform MD nodes (Womersley, 2020); convective derivatives were approximated using stencils of size $n = 74$ based on Gaussian RBFs with $\varepsilon = 8$, without supplementary polynomials and, as hyperviscosity, the Laplacian to eighth power. Figure 5.13(a) shows the DM eigenvalues in a smaller case of $N = 3,600$, $n = 17$, and no hyperviscosity. The vast number of eigenvalues in the RHP makes this case hopelessly unstable already for very short times. Part (b) of this figure shows the eigenvalues after a small amount of Laplacian-squared hyperviscosity has been applied. As all the "offending" eigenmodes are highly oscillatory,[29] the hyperviscosity has shifted all of these to the LHP, where they naturally decay during time stepping. The physically relevant ones have been left almost perfectly intact along a central part of the imaginary axis.[30]

While hyperviscosity has been highly successful in many applications, the tuning of its parameters (especially the factor γ) remains a somewhat delicate task, in spite of available guidelines. One recent approach that reduces (or in some cases perhaps eliminates) the need for hyperviscosity stabilization is described in Section 5.4.5.

5.4.4 Time Stepping Stability Conditions

Time stepping of RBF–FD discretizations is virtually always carried out using the MOL (method of lines) approach. Although PDEs typically have spatially and temporarily variable coefficients (and often are nonlinear), one usually obtains accurate stability information by verifying that the eigenvalues λ_i of the DM (or of its linearized version) fall within the complex plane region covered by $\frac{1}{k}$ times the ODE solver's stability domain in the ξ-plane, as visualized for several ODE solvers in Figure 3.5. Since, for every ODE solver, the edge of the stability domain has a vertical tangent at $\xi = 0$ (with the region immediately to the right of the origin always outside the domain), no solver will be stable in the case seen in Figure 5.13(a), whereas all the solvers illustrated in Figure 3.5 will work for the case in Figure 5.13(b), as long as the time step k is sufficiently small.[31]

[29] or else they would have to be physically relevant, and then not have eigenvalues in the RHP.
[30] for this test problem with their imaginary parts correctly at integer levels.
[31] There are important ODE solvers that do not allow any eigenvalues that are on either side of the imaginary axis. One such example is leapfrog, as described in Example 4.1.7.

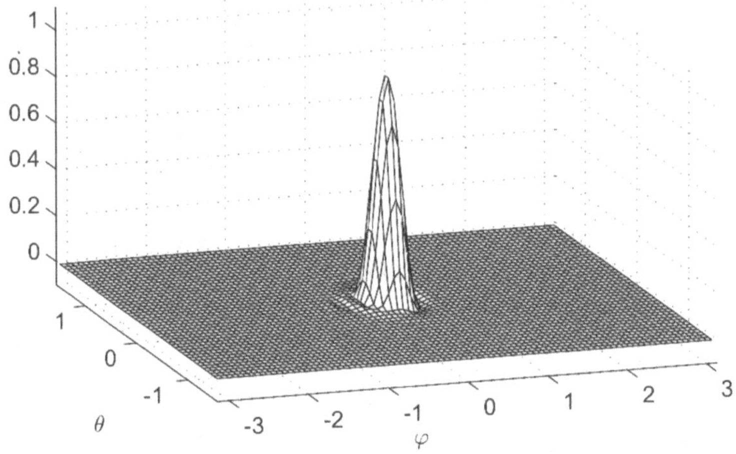

Figure 5.12 Numerical solution of a cosine bell after time stepping corresponding to 1,000 revolutions around a sphere, as described in Section 5.4.3. Only small errors are visible near its base, partly caused by the cosine bell there not being twice differentiable. The loss in peak height is not graphically discernible, and there are no signs of trailing waves along its path (angled 45° to the coordinate system's equatorial plane $\theta = 0$). In this display, φ and θ correspond to longitude and latitude, respectively, on the sphere. Compared to the 1-D convected pulse in Figure 4.6, the bell here (with $N = 25,600$ nodes on the sphere) is about twice as finely resolved and has traveled over 200 times further (in terms of average node spacings). Figure reproduced from Fornberg and Lehto (2011).

5.4.5 Staggered RBF–FD Approximations

The examples of staggered variables given in Section 4.1.4.2 were particularly favorable to this procedure due to the specific terms the PDEs contained (and did not contain). Discretizations of the incompressible Navier–Stokes (NS) equations can also benefit from staggering, as described for 3-D already in Harlow and Welch (1965). Focusing for simplicity here on the 2-D case, these equations can be written nondimensionalized as

$$\begin{aligned}
\frac{\partial u}{\partial t} &= -\left(u\frac{\partial u}{\partial x} + v\frac{\partial u}{\partial y}\right) - \frac{\partial p}{\partial x} + \frac{1}{\text{Re}}\left(\frac{\partial^2 u}{\partial x^2} + \frac{\partial^2 u}{\partial y^2}\right), \\
\frac{\partial v}{\partial t} &= -\left(u\frac{\partial v}{\partial x} + v\frac{\partial v}{\partial y}\right) - \frac{\partial p}{\partial y} + \frac{1}{\text{Re}}\left(\frac{\partial^2 v}{\partial x^2} + \frac{\partial^2 v}{\partial y^2}\right), \\
\frac{\partial u}{\partial x} &+ \frac{\partial v}{\partial y} = 0,
\end{aligned} \tag{5.15}$$

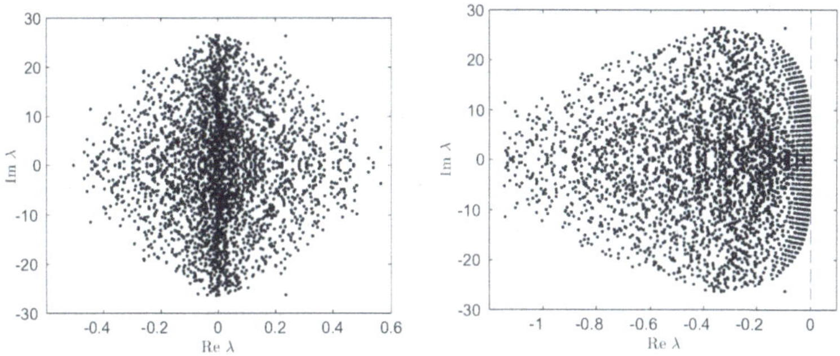

(a) Without hyperviscosity (b) With hyperviscosity

Figure 5.13 Eigenvalues of the differentiation matrix (DM) for the convection around a sphere test problem. Figure reproduced from Fornberg and Lehto (2011).

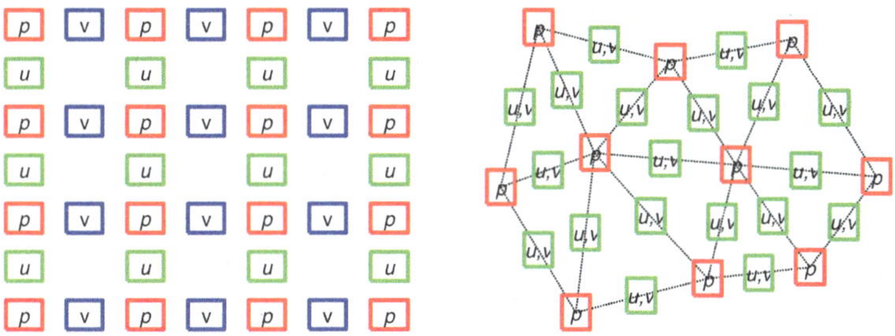

(a) Cartesian grid (b) Quasi-uniform nodes

Figure 5.14 Examples of staggering suitable for the 2-D Navier–Stokes equations.

commonly staggered as shown in Figure 5.14(a). It is reported in Chu and Schmidt (2023) that the staggering conceptually illustrated in Figure 5.14(b) greatly reduces (or even entirely eliminates) the need for applying stabilizing hyperviscosity, with RBF–FD PHS+poly illustrations provided that include also very high Reynolds number (Re) cases (with the flows then highly advection dominated). Here, the p-nodes are distributed quasi-uniformly. The connection lines between adjacent p-nodes can be created directly by a node generator (such as `distmesh`, Persson and Strang, 2004), or separately by a *Delaunay triangulation*. Both u and v are here represented at the halfway points between the p-nodes (rather than at separate locations, as in Figure 5.14(a) and in typical finite volume discretizations).

Before leaving the topic of FD-inspired RBF–FD enhancements, we note that also the concept of compact stencils for Poisson-type PDEs (cf. Section 4.3.1.1) can be carried over to RBF–FD discretizations (Wright and Fornberg, 2006).

5.4.6 RBF–FD and Boundaries

5.4.6.1 Fixed Boundaries

Boundaries restrict the physical domain of interest, while interfaces separate regions with different governing equations, in both cases supplemented by boundary or interface conditions. Many methods developed for grid-based FD schemes carry over, with the additional benefit that with RBF–FD one can also take advantage of the freedom to place nodes to align with (or to straddle) such features.[32] In both cases, one option is to use one-sided approximations on each side, utilizing the fact that Runge-phenomenon-type inaccuracies tend to get reduced with increasing RBF–FD stencil sizes, as described in Section F.3 and Bayona (2019b); Bayona et al. (2019).[33] Another possibility is to create weight sets that remain accurate also for stencils that cross over a curved interface. One approach for this is based on the observation that for smooth solutions, using RBF–FD PHS+poly, the RBF part fades out under refinement. With typically linear interface conditions, it may suffice to "kink" the supplementary polynomials in a way that, to leading orders, accounts for curved interfaces. This was used for time-dependent elastic waves in Martin and Fornberg (2017a), and for steady-state heat conduction in Martin and Fornberg (2017b) (successfully also in cases with interfaces located much closer to each other than the node spacing).

5.4.6.2 Moving Boundaries

There is much current research in progress on RBF–FD for moving boundaries and interfaces. For example, for free surface flows, options that have been considered include dynamically moving a thin layer of nodes near to the surface, or keeping nodes fixed while implementing boundary conditions via constraint equations, see Section D.3. The approach in Shankar et al. (2021) applies to convection–diffusion problems and combines a semi-Lagrangian formulation with rapid node set modification and high-accuracy time stepping.

5.4.7 RBF–FD on Surfaces

Particle and finite element methods rarely reach high orders of accuracy. The main RBF–FD related approaches for solving PDEs on surfaces include local transforms to tangent planes (Gunderman et al., 2020; Wright et al., 2023) and

[32] In the case of solving elliptic PDEs, such alignment may be unnecessary (Nielsen and Fornberg, 2024).

[33] This observation led to the development of the enhanced Gregory quadrature method described in Section F.2.

the closest point method[34] (März and Macdonald, 2012; Petras et al., 2019; Ruuth and Merriman, 2008). For evolving surfaces, several approaches are compared in Wendland and Künemund (2020) (where it is remarked that RBF–FD implementations are easy to use, but still lack solid theoretical foundations).

5.4.8 RBF–FD for Time-Independent Equations

It was observed in Section 5.4.1 for RBF–FD interpolation based on PHS+poly that

i. the formal order of accuracy/convergence rate will be determined by the polynomial degrees only (i.e., one easily realizes quite high orders, such 4, 5, or 6).

This applies also in the context of solving PDEs, including of equilibrium (elliptic) type (Bayona et al., 2017). It is also important that PHS+poly-type RBF–FD stencils, without any special treatment, much reduce the effects of Runge phenomena at domain edges.[35]

Key features include

i. Much reduced need to use ghost nodes or similar techniques to handle boundaries, see Fedoseyev et al. (2002); Larsson and Fornberg (2003); Smith (1965) and Example F.3.1.
ii. High-order accuracy can be realized everywhere across the domain, including near the boundaries.
iii. Favorable numerical stability for the resulting large sparse linear systems.

The large linear systems that arise with both regular FD and RBF–FD methods can typically be solved effectively with standard iterative methods[36] including multigrid or (for RBF–FD) multilevel methods, see Sections 4.3.5 and G.2.4 and also Lawrence et al. (2024); Wright et al. (2023); Zamolo et al. (2019).

The literature on RBF–FD methodologies and applications is rapidly evolving, and the descriptions stated earlier are mainly intended to give some historically significant pointers.

[34] This method is closely related to the orthogonal gradient (OGr) method, extending the PDE to an embedding space near to the surface and in this space construct and then restrict differential operators.

[35] As noted in Section 5.4.6, and described further in Section D.3.

[36] For RBF–FD, see Leinen (2024).

6

FD in the Complex Plane

Number systems originated with positive integers. Subsequent extensions include rational and irrational numbers, zero and negative numbers, followed by complex numbers (with still more types considered[1]). If not already before, the connection between numbers and counting objects vanished with the introduction of negative numbers (simplifying the rules for subtraction). The great utility of complex numbers ($z = x + iy$ with x and y real, and $i = \sqrt{-1}$) lies similarly in how these significantly simplify (and provide powerful shortcuts to) many mathematical tasks that, in their formulation and their end results, only involve real-valued quantities. Jacques Hadamard (paraphrasing an earlier statement in 1900 by Paul Painlevé) wrote, "The shortest path between two truths in the real domain passes through the complex domain." With most computational tasks forming a link between physically motivated real-valued input and output quantities,[2] it is natural to ask if this "shortest path" at times can benefit from FD methods in the complex plane.

There are numerous textbooks on complex variables and analytic functions, for example, Ahlfors (1966); Brown and Churchill (2013); Fornberg and Piret (2020); Wegert (2012) (with the latter two rich in function illustrations). We refer to these (or similar) books for much more background than what is given in Section 6.1.

[1] Generalizations beyond complex, such as quaternions, come however with severe limitations typically including multiplication no longer being commutative.

[2] There are in quantum mechanics cases where complex numbers have a direct physical meaning (Renou et al., 2023).

6.1 Analytic Functions

The standard and special functions of the applied sciences can be Taylor expanded around most points x_0:

$$f(x) = \sum_{k=0}^{\infty} a_k (x - x_0)^k, \tag{6.1}$$

converging for some nonzero range $|x - x_0| < R$. When this is the case, simply changing $x \longrightarrow z$ (complex) (and renaming $x_0 \longrightarrow z_0$) defines the unique analytic extension of $f(x)$ to $|z - z_0| < R$.

Analyticity is also often described from a different starting point: $f(z)$ is analytic at a (complex) location z if

$$f'(z) = \lim_{\Delta z \to 0} \frac{f(z + \Delta z) - f(z)}{\Delta z} \tag{6.2}$$

is finite and takes the same value independently of from which direction in the complex plane $\Delta z \longrightarrow 0$. From either (6.1) or (6.2) follow the *Cauchy–Riemann* (CR) *equations*. Separating in real and imaginary parts

$$f(z) = f(x + iy) = u(x, y) + i\,v(x, y) \tag{6.3}$$

leads to

$$\begin{cases} \dfrac{\partial u}{\partial x} = \dfrac{\partial v}{\partial y}, \\ \dfrac{\partial v}{\partial x} = -\dfrac{\partial u}{\partial y}. \end{cases} \tag{6.4}$$

These CR equations impose constraints on $f(z)$ compared to all possible complex functions of a complex variable, in exchange for far-reaching analytical and computational opportunities. As already mentioned, virtually all standard and special functions "naturally" obey these relations.

6.2 Introductory Comments on Differentiation in the Complex Plane

Before turning our attention to FD approximations on equispaced grids in the complex plane, we illustrate graphically a key benefit of complex plane evaluations and mention two early approaches for complex plane-enhanced numerical differentiation.

By (6.1) and the relation

$$a_k = \frac{f^{(k)}(x_0)}{k!}, \tag{6.5}$$

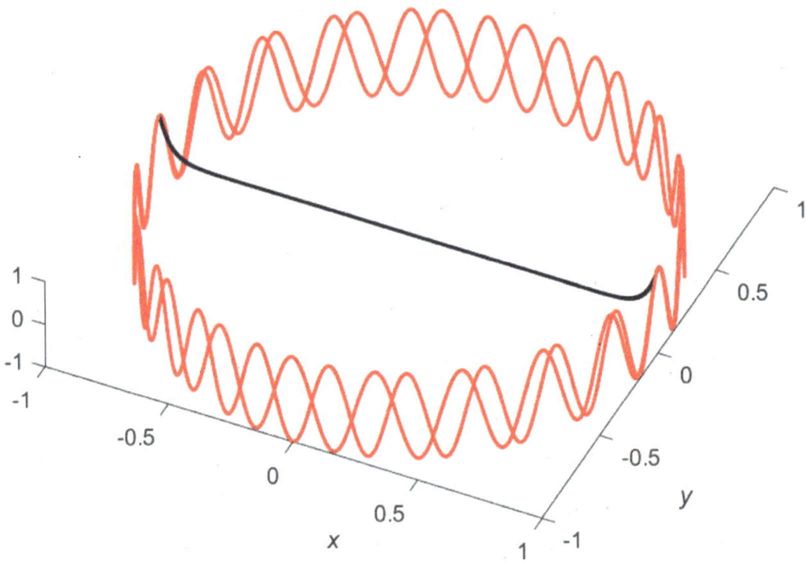

Figure 6.1 The functions x^{20} and x^{22} along $-1 \leq x \leq 1$ (two black curves, visually indistinguishable from each other) and the real parts of z^{20} and z^{22} around the unit circle (two red curves).

numerical differentiation is equivalent to numerical evaluation of Taylor coefficients. For the illustration in Figure 6.1, we set $x_0 = 0$ and write z in place of x. The figure shows that the Taylor basis functions x^k for higher k become virtually indistinguishable from each other along a real axis interval (here $[-1, 1]$), while the functions z^k become orthogonal around the unit circle – making the problem of numerically determining Taylor coefficients from equispaced data then vastly better conditioned.

6.2.1 Complex Step Method

Following Squire and Trapp (1998), let $f(x)$ be an analytic function that is real valued for x real, with the task to approximate $f'(x)$ at a real x-point. Straightforward FD2

$$f'(x) \approx \frac{1}{2} \left(f(x+h) - f(x-h) \right) / h$$

will (in standard double precision) give an error as a function of h following the FD2 curve in Figure 1.2. However, by the first line in (6.4), we can just as well use the approximation

$$f'(x) \approx \frac{1}{2}\left(\mathrm{Im}f(x+ih) - \mathrm{Im}f(x-ih)\right)/h. \tag{6.6}$$

In contrast to $\mathrm{Re}f(x\pm h)$, $\mathrm{Im}f(x\pm ih)$ goes to zero for small h as $O(h)$. Assuming the code for $f(x)$ complex gives full significant digits for $\mathrm{Im}f(x\pm ih)$, the typical $O(10^{-16})/h$ rounding error term becomes of size $O(h\cdot 10^{-16})/h = O(10^{-16})$. With only the $O(h^2)$ truncation error then left, it becomes safe to reduce h to below 10^{-8}, resulting in full machine precision accuracy.[3]

Looking again at Figure 1.2, we see that FD8 for approximating $f'(x)$ would reach an error of about 10^{-14} if applied with $h \approx 10^{-1}$ – good but not quite as good as 10^{-16}.

6.2.2 Numerical Differentiation Using Cauchy's Integral Formula

We recall from Section 1.3 that, when approximating the kth derivative with an FD formula of accuracy order p, the expected error is the sum of truncation error $O(h^p)$ and rounding error $O(10^{-16})/h^k$. If k is large (say, 20 or 50), this rounding error is disastrous.

For an analytic function $f(z)$, Cauchy's integral theorem gives the kth derivative $f^{(k)}(z)$ as the integral

$$f^{(k)}(z) = \frac{k!}{2\pi i}\oint_\Gamma \frac{f(\xi)}{(\xi - z)^{k+1}}\,d\xi. \tag{6.7}$$

The contour Γ goes around z once in the positive direction, without enclosing any singularities of $f(z)$. Typically, Γ is chosen as a circle centered at z with some radius r. The fundamental advantage offered by (6.7) is that this formula is exact, with no need to make r (the counterpart to h) small in order to bring down a truncation error. The integrand is periodic, and the trapezoidal rule (discussed further in Chapters 7 and 8), is spectrally accurate.[4] Using as large an r as the features of $f(z)$ allow, neither truncation errors nor rounding errors are problematic, and close to machine accuracy is reachable, even for derivatives of very high orders. This feature was first explored in Lyness and Moler (1967); Lyness and Sande (1971), with the enhancement in Fornberg (1981) of automating the choice of r. Further perspectives on utilizing data around a circle in the complex plane can be found in Austin et al. (2014).

[3] An alternative way to motivate it is to consider the one-sided approximation $f'(x) \approx (\mathrm{Im}f(x+ih) - \mathrm{Im}f(x))/h = \mathrm{Im}f(x+ih)/h$, which is free from canceling subtraction. The complex step method is generalized and applied to wave equations in Abreu et al. (2015).

[4] If results are desired for many k-values, these can be obtained efficiently with the FFT algorithm (Appendix C.4).

6.2.3 Connection between Taylor and Fourier Series Expansions

The Cauchy integral formula approach can also be arrived at from a different starting point. From (6.1) and (6.5) (and visually apparent in Figure 6.1), a Taylor expansion takes the form of a Fourier expansion when inspected around the periphery of a circle with the same center as the Taylor expansion. This allows the Taylor coefficients (i.e., the derivatives at the center point) to be computed by applying the (inverse) FFT to the function values around the circle. *Aliasing* effects (different Fourier modes looking the same at equispaced sampling points, see Section C.3.2) need to be compensated for (which can be combined with extrapolation and a search for a good radius; see Bornemann, 2011, Fornberg, 1981).

The discussion above expresses the fact that numerical approximation of high-order derivatives is an ill-conditioned problem along the real axis but (for analytic functions) becomes well conditioned if function data is also picked up off the real axis (much as is the case when numerically evaluating inverse Laplace transforms).[5] Grid-based complex plane FD methods are often more practical than using data around circles and are discussed in the remainder of this chapter.

6.3 FD Stencils in the Complex Plane for Analytic Functions

The methods for calculating FD weights described in Sections 1.2.1.1–1.2.1.4 work just as well for nodes z_k in the complex plane as for nodes x_k along the real axis. Table 6.1 shows the weights for the first four derivatives in the two smallest centered square stencil cases (denoted $n = 1$ and $n = 2$; with total number of nodes $N = (2n + 1)^2$).[6] Two features are notable:

i. The orders of accuracy are high even with small stencil sizes.
ii. For larger stencils, the magnitude of the weights decreases rapidly with the distance from the stencil center (described in more detail in Section 6.3.1).

If we seek FD formulas with all coefficients real valued (applicable to 2-D harmonic functions with matching imaginary parts not readily available), the accuracy orders typically become significantly reduced.[7] For example, the

[5] See the equation in Fornberg and Piret (2020), top of page 247, versus approximating the Bromwich integral.

[6] One-sided (or partly one-sided) approximations can be constructed similarly, for use at or near domain boundaries.

[7] The 3×3 stencil for the fourth derivative in Table 6.1 being an exception.

Table 6.1 *FD weights for the first four derivatives in the two smallest square stencil cases. For these first four derivatives, all the shown* 3×3 *stencils are accurate to* $O(h^8)$ *and the* 5×5 *stencils to* $O(h^{24})$.

	$n = 1$; Stencil size 3×3	$n = 2$; Stencil size 5×5
f'	$\frac{1}{40h}\begin{bmatrix} -1-i & -8i & 1-i \\ -8 & 0 & 8 \\ -1+i & 8i & 1+i \end{bmatrix} f$	$\frac{1}{39h}\begin{bmatrix} \frac{1+i}{12240} & \frac{4(-1-i)}{765} & \frac{i}{34} & \frac{4(1-i)}{765} & \frac{-1+i}{12240} \\ \frac{4(-1-i)}{765} & \frac{8(-1-i)}{9} & & \frac{8(1-i)}{9} & \frac{4(1-i)}{765} \\ \frac{1}{34} & -8 & 0 & 8 & \frac{-1}{34} \\ \frac{4(-1+i)}{765} & \frac{8(-1+i)}{9} & 8i & \frac{8(1+i)}{9} & \frac{4(1+i)}{765} \\ \frac{1-i}{12240} & \frac{4(-1+i)}{765} & \frac{-i}{34} & \frac{4(1+i)}{765} & \frac{-1-i}{12240} \end{bmatrix} f$
f''	$\frac{1}{20h^2}\begin{bmatrix} i & -8 & -i \\ 8 & 0 & 8 \\ -i & -8 & i \end{bmatrix} f$	$\frac{1}{39h^2}\begin{bmatrix} \frac{-i}{12240} & \frac{8(-1+3i)}{3825} & \frac{1}{34} & \frac{8(-1-3i)}{3825} & \frac{i}{12240} \\ \frac{8(1+3i)}{3825} & \frac{16i}{9} & -16 & \frac{-16i}{9} & \frac{8(1-3i)}{3825} \\ \frac{-1}{34} & 16 & 0 & 16 & \frac{-1}{34} \\ \frac{8(1-3i)}{3825} & \frac{-16i}{9} & -16 & \frac{16i}{9} & \frac{8(1+3i)}{3825} \\ \frac{i}{12240} & \frac{8(-1-3i)}{3825} & \frac{1}{34} & \frac{8(-1+3i)}{3825} & \frac{-i}{12240} \end{bmatrix} f$
$f^{(3)}$	$\frac{3}{40h^3}\begin{bmatrix} 1-i & 16i & -1-i \\ -16 & 0 & 16 \\ 1+i & -16i & -1+i \end{bmatrix} f$	$\frac{1}{13h^3}\begin{bmatrix} \frac{-1+i}{48960} & \frac{8(7-i)}{19125} & \frac{-i}{68} & \frac{8(-7-i)}{19125} & \frac{1+i}{48960} \\ \frac{8(1-7i)}{19125} & \frac{8(1-i)}{9} & 16i & \frac{8(-1-i)}{9} & \frac{8(-1-7i)}{19125} \\ \frac{1}{68} & -16 & 0 & 16 & \frac{-1}{68} \\ \frac{8(1+7i)}{19125} & \frac{8(1+i)}{9} & -16i & \frac{8(-1+i)}{9} & \frac{8(-1+7i)}{19125} \\ \frac{-1-i}{48960} & \frac{8(7+i)}{19125} & \frac{i}{68} & \frac{8(-7+i)}{19125} & \frac{1-i}{48960} \end{bmatrix} f$
$f^{(4)}$	$\frac{3}{10h^4}\begin{bmatrix} -1 & 16 & -1 \\ 16 & -60 & 16 \\ -1 & 16 & -1 \end{bmatrix} f$	$\frac{1}{13h^4}\begin{bmatrix} \frac{1}{24480} & \frac{32(-9-13i)}{95625} & \frac{-1}{34} & \frac{32(-9+13i)}{95625} & \frac{1}{24480} \\ \frac{32(-9+13i)}{95625} & \frac{-32}{9} & 64 & \frac{-32}{9} & \frac{32(-9-13i)}{95625} \\ \frac{-1}{34} & 64 & \frac{-1208181}{5000} & 64 & \frac{-1}{34} \\ \frac{32(-9-13i)}{95625} & \frac{-32}{9} & 64 & \frac{-32}{9} & \frac{32(-9+13i)}{95625} \\ \frac{1}{24480} & \frac{32(-9+13i)}{95625} & \frac{-1}{34} & \frac{32(-9-13i)}{95625} & \frac{1}{24480} \end{bmatrix} f$

highest accuracy orders then possible for approximating the first derivative are for 3×3 stencils realized by

$$\frac{\partial f}{\partial x} = \frac{1}{12h}\begin{bmatrix} -1 & 0 & 1 \\ -4 & 0 & 4 \\ -1 & 0 & 1 \end{bmatrix} f + O(h^4) \tag{6.8}$$

and for 5×5 stencils by

$$\frac{\partial f}{\partial x} = \frac{1}{77h}\begin{bmatrix} -\frac{37}{120} & -\frac{8}{5} & 0 & \frac{8}{5} & \frac{37}{120} \\ -\frac{34}{15} & -2 & 0 & 2 & \frac{34}{15} \\ -\frac{23}{6} & -\frac{40}{3} & 0 & \frac{40}{3} & \frac{23}{6} \\ -\frac{34}{15} & -2 & 0 & 2 & \frac{34}{15} \\ -\frac{37}{120} & -\frac{8}{5} & 0 & \frac{8}{5} & \frac{37}{120} \end{bmatrix} f + O(h^{12}). \qquad (6.9)$$

6.3.1 Magnitude of Weights as Function of Distance to Stencil Center

The bottom line in Tables 1.1 and 1.2 showed that weights at $x_k = k$ (k an integer) decay in magnitude as $O(1/|k|)$ and $O(1/|k^2|)$, respectively (then alternating between these two cases for higher odd and even derivative orders). In the complex FD case, with nodes at $z_k = \mu + i\nu$ (μ, ν integers), there is again an algebraic decay present, but this is now multiplied by the extremely rapidly decay factor $e^{-\frac{\pi}{2}(\mu^2+\nu^2)}$ (first noted and proved in Fornberg, 2022). Figures 6.2 and 6.3 illustrate this coefficient decay in two different ways.

Figure 6.2 shows the magnitude of the weights in a central 5×5 area on a unit-spaced grid when approximating the fourth derivative[8] at the origin ($\mu = \nu = 0$) in the cases of $n = 1, 2, 3$ (i.e., 3×3, 5×5, and 7×7 size stencils). The weights are seen to converge rapidly.

Figure 6.3 shows, in the same case (but now also with $n = 4$), the logarithm of magnitude of the weights. The parabolic shape in this log-linear plot corresponds to the weights decaying not just algebraically with μ and ν, but furthermore with the factor $e^{-\frac{\pi}{2}(\mu^2+\nu^2)}$ (holding for derivatives of any order).

Consequences of this decay include:

i. The PS limit of increasing orders of accuracy not only exists but remains highly local (as is the mathematical concept of a derivative).
ii. Complex plane FD stencils can be applied very close to singularities – as will be utilized for example in Section 7.4.
iii. For a given accuracy order p, the error constant in the $O(h^p)$ term is much smaller in the complex case.

The last of these issues is illustrated in Figure 6.4, which shows the errors (computed in extended precision arithmetic) when approximating $\left.\frac{d^4}{dx^4}e^{2x}\right|_{x=0}$ with 25-node stencils extending along the real axis and in the complex plane,

[8] The lowest-order derivative for which the weight at the center node is nonzero, see Table 6.1.

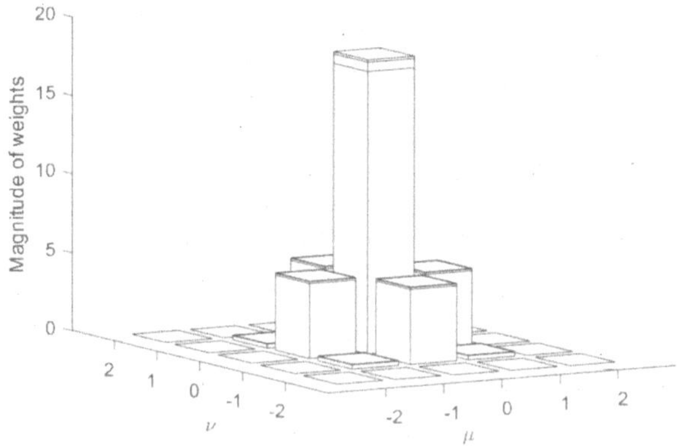

Figure 6.2 Magnitude of weights when approximating the fourth derivative at the origin. Central 5×5 region shown for stencil sizes 3×3, 5×5, and 7×7 (illustrating both strong locality and rapid convergence to a PS limit).

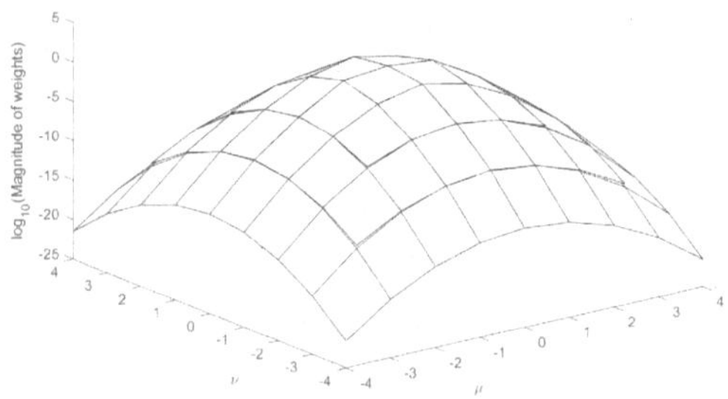

Figure 6.3 $\log_{10}$(Magnitude of weights) when approximating the fourth derivative at the origin in the $n = 1, 2, 3, 4$ cases (stencil sizes from 3×3 to 9×9). The paraboloid-shaped surface in this log-linear plot confirms the $e^{-\frac{\pi}{2}(\mu^2+\nu^2)}$ (Gaussian-type) decay of the weights. The differences in the magnitude of the weights between the four cases are barely visible right at the stencil corners.

respectively. For the same formal order of accuracy (seen as the slope of the lines), the error constant in the latter case is smaller by a factor of about 10^{-15}.

6.3.2 PS Limit for Complex Plane FD Formulas

Some analysis of this PS limit is given in Fornberg (2023). Table 6.2 gives the weights in the central 7×7 area for the PS stencils for the first and second derivative. Along the real and imaginary axes, we see, *before the factor* $e^{-\frac{\pi}{2}(\mu^2+\nu^2)}$

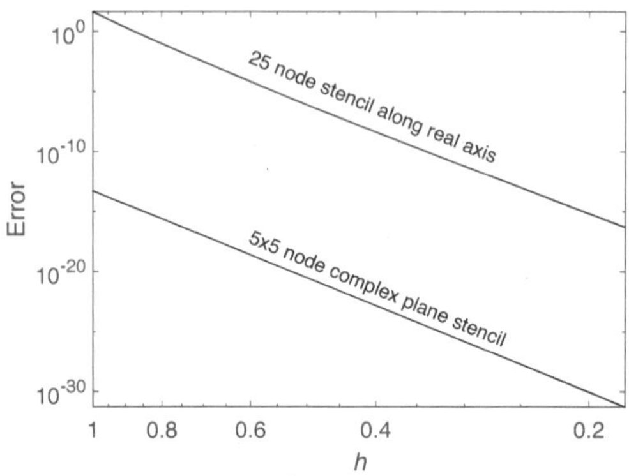

Figure 6.4 Comparison of errors in approximating $\left.\frac{d^4}{dx^4}e^{2x}\right|_{x=0}$.

has been applied, rational coefficients with decay rates that are reminiscent of the real-valued PS cases. The complex plane PS weights for interpolation to an arbitrary in-between grid point location feature a similar Gaussian decay, and can be given explicitly following a single evaluation (for each interpolation location) of the Weierstrass σ-function[9] – see Fornberg (2023), Section 5.2.

Figure 6.5 shows contour lines of this factor $e^{-\frac{\pi}{2}(\mu^2+\nu^2)}$ on a unit-spaced grid corresponding to the accuracies of single, double, and quad precision arithmetic. The double precision level of 10^{-16} is exceeded in only 69 grid points.

6.3.3 Characteristic Function for an FD Approximation

Given an FD approximation

$$f^{(k)}(0) = \frac{1}{h^k}\sum_{j=1}^{N} w_j f(h z_j) + O(h^p), \tag{6.10}$$

an associated *characteristic function* was introduced in Wegert (2024)

$$w(z) = \sum_{j=1}^{N} \frac{w_j}{z - z_j}. \tag{6.11}$$

[9] Not necessary in the special case of interpolating to a grid midpoint location.

Table 6.2 *Central 7×7 area of the PS stencils for the first two derivatives. The node locations are $z_k = \mu + i\,\nu$, with μ and ν integers. Each of the numbers shown should be multiplied by $e^{-\frac{\pi}{2}(\mu^2+\nu^2)}$ to obtain the actual PS weight.*

First derivative

$$\frac{1}{h}\begin{bmatrix} & \vdots & \vdots & \vdots & \vdots & \vdots & \vdots & \vdots & \\ \cdots & \frac{-1-i}{6} & \frac{-2-3i}{13} & \frac{-1-3i}{10} & \frac{-i}{3} & \frac{1-3i}{10} & \frac{2-3i}{13} & \frac{1-i}{6} & \cdots \\ \cdots & \frac{-3-2i}{13} & \frac{1+i}{4} & \frac{-1-2i}{5} & \frac{i}{2} & \frac{1-2i}{5} & \frac{-1+i}{4} & \frac{3-2i}{13} & \cdots \\ \cdots & \frac{-3-i}{10} & \frac{-2-i}{5} & \frac{-1-i}{2} & -i & \frac{1-i}{2} & \frac{2-i}{5} & \frac{3-i}{10} & \cdots \\ \cdots & \frac{-1}{3} & \frac{1}{2} & -1 & 0 & 1 & \frac{-1}{2} & \frac{1}{3} & \cdots \\ \cdots & \frac{-3+i}{10} & \frac{-2+i}{5} & \frac{-1+i}{2} & i & \frac{1+i}{2} & \frac{2+i}{5} & \frac{3+i}{10} & \cdots \\ \cdots & \frac{-3+2i}{13} & \frac{1-i}{4} & \frac{-1+2i}{5} & \frac{-i}{2} & \frac{1+2i}{5} & \frac{-1-i}{4} & \frac{3+2i}{13} & \cdots \\ \cdots & \frac{-1+i}{6} & \frac{-2+3i}{13} & \frac{-1+3i}{10} & \frac{i}{3} & \frac{1+3i}{10} & \frac{2+3i}{13} & \frac{1+i}{6} & \cdots \\ & \vdots & \vdots & \vdots & \vdots & \vdots & \vdots & \vdots & \end{bmatrix}$$

Second derivative

$$\frac{1}{h^2}\begin{bmatrix} & \vdots & \vdots & \vdots & \vdots & \vdots & \vdots & \vdots & \\ \cdots & \frac{i}{9} & \frac{-10+24i}{169} & \frac{-4+3i}{25} & \frac{-2}{9} & \frac{-4-3i}{25} & \frac{-10-24i}{169} & \frac{-i}{9} & \cdots \\ \cdots & \frac{10+24i}{169} & \frac{-i}{4} & \frac{-6+8i}{25} & \frac{1}{2} & \frac{-6-8i}{25} & \frac{i}{4} & \frac{10-24i}{169} & \cdots \\ \cdots & \frac{4+3i}{25} & \frac{6+8i}{25} & i & -2 & -i & \frac{6-8i}{25} & \frac{4-3i}{25} & \cdots \\ \cdots & \frac{2}{9} & \frac{-1}{2} & 2 & 0 & 2 & \frac{-1}{2} & \frac{2}{9} & \cdots \\ \cdots & \frac{4-3i}{25} & \frac{6-8i}{25} & -i & -2 & i & \frac{6+8i}{25} & \frac{4+3i}{25} & \cdots \\ \cdots & \frac{10-24i}{169} & \frac{i}{4} & \frac{-6-8i}{25} & \frac{1}{2} & \frac{-6+8i}{25} & \frac{-i}{4} & \frac{10+24i}{169} & \cdots \\ \cdots & \frac{-i}{9} & \frac{-10-24i}{169} & \frac{-4-3i}{25} & \frac{-2}{9} & \frac{-4+3i}{25} & \frac{-10+24i}{169} & \frac{i}{9} & \cdots \\ & \vdots & \vdots & \vdots & \vdots & \vdots & \vdots & \vdots & \end{bmatrix}$$

The two leading terms for $w(z)$ as $z \to \infty$ provide the values for both k and p since[10]

$$w(z) = \frac{k!}{z^{k+1}} + O\left(\frac{1}{z^{k+p+1}}\right). \tag{6.12}$$

A follow-up result to this (also shown in Wegert (2024)) is that, if $w_j\big|_k$ denote the weights at nodes z_j (which need not be grid based) for an approximation to $f^{(k)}(0)$, one can obtain an approximation to $f^{(k-1)}(0)$ of equal accuracy order

[10] Assuming for both (6.10) and (6.12) that the powers in the O-symbols are sharp.

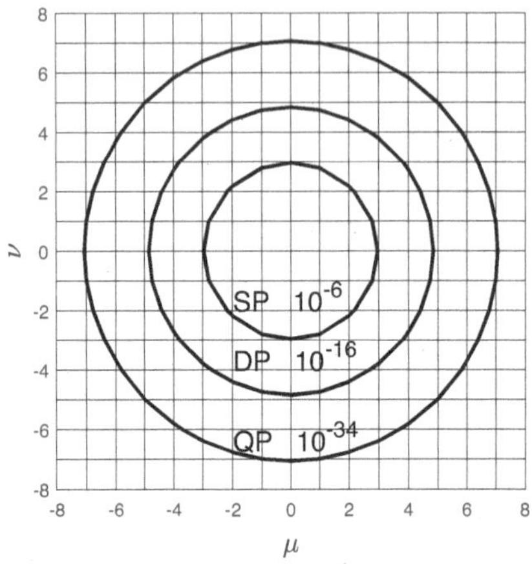

Figure 6.5 Contour lines of the factor $e^{-\frac{\pi}{2}(\mu^2+\nu^2)}$ corresponding to single (SP), double (DP), and quad (QP) precisions.

by

$$w_j\big|_{k-1} = \frac{z_j}{k}\left(w_j\big|_k\right), \quad j = 1, 2, \ldots, N, \quad k > 1. \tag{6.13}$$

With this relation, the weights in Table 6.1 for $f^{(3)}$, $f^{(2)}$, $f^{(1)}$ follow from the ones for $f^{(4)}$, and in Table 6.2 the first-derivative approximation from the second derivative one. Also, Table 1.1 follows from Table 1.2 and Table 4.3 from Table 4.1. These examples of applying (6.13) are somewhat exceptional, as one usually (for a fixed node set z_j and in the absence of symmetries) would expect to be able to gain an order of accuracy each time k is reduced by 1.

6.4 Connections between Analytic Functions and Laplace's Equation in 2-D

6.4.1 Conformal Mappings

Given an analytic function $f(z)$, equation (6.3) can be seen as a 2-D mapping $(x, y) \rightarrow (u, v)$, *conformal* (locally angle-preserving) if $f'(z) \neq 0$ within some domain D. In the (u, v)-space, the domain becomes different – D'. From the CR equations (6.4) follow the fact that if some function satisfies Laplace's equation in (x, y) within D, it will do so also in (u, v) within D'. If it is practical to map to a region D' of particularly simple shape, this can sometimes provide analytic solution opportunities and, at other times, simplify application of FD (or other numerical) methods. Discussions of conformal mappings can be found in most complex variables textbooks.[11] In contemporary computing, the role of conformal mappings is diminishing in view of more flexible grid generation schemes, see for example Thompson et al. (1998).

6.4.2 Connection to Compact Stencils

The FD discussion in Section 4.3.1 applied in any number of dimensions. In the special case of 2-D, analytic functions provide special opportunities as a result of the CR equations (6.4), which imply that both the real part $u(x, y)$ and the imaginary part $v(x, y)$ of an analytic function satisfies Laplace's equation, that is,

$$\frac{\partial^2 u}{\partial x^2} + \frac{\partial^2 u}{\partial y^2} = 0. \tag{6.14}$$

Continuing the sequence of stencils in the left column of Table 6.1 to the eighth derivative will show

$$f^{(8)}(0) = \frac{504}{h^8} \begin{bmatrix} 1 & 4 & 1 \\ 4 & -20 & 4 \\ 1 & 4 & 1 \end{bmatrix} f + O(h^4). \tag{6.15}$$

Separating real and imaginary parts, and multiplying by $\frac{h^6}{6 \cdot 504}$, we obtain for example for $u(x, y)$,

$$\begin{bmatrix} 1 & 4 & 1 \\ 4 & -20 & 4 \\ 1 & 4 & 1 \end{bmatrix} \frac{u}{6h^2} = 0 + O(h^6), \tag{6.16}$$

in agreement with the comment following (4.23).[12]

[11] See, for example, Chapter 6 in Ahlfors (1966) or Chapter 8 in Fornberg and Piret (2020); or in more detail Nehari (1995 (originally published by Mc-Graw Hill 1952)).

[12] The 5×5 counterpart to the matrix in (6.16), approximating the 24th derivative, is less directly applicable, as eight of its entries are complex.

7

FD-based Methods for Quadrature and Infinite Sums

Derivatives of orders $1, 2, 3, \ldots$ (as well as of order 0 for interpolation) are all spatially local operators, and thus well suited for FD approximations (also local). Quadrature can be thought of as of order -1, while fractional derivatives (discussed in Chapter 8) are of noninteger orders. As these latter two cases do not correspond to local operators, this and Chapter 8 may at first seem to fall outside the topic of this book. However, in both of these cases, it transpires that the simple trapezoidal rule (TR) becomes highly effective when enhanced with local FD-type corrections at each interval end.

We will not be discussing Gaussian quadrature methods, as these have no similar FD connections, and also require highly specialized node sets. In applications, numerical quadrature is rarely a sole goal in itself, but more often an add-on step in some larger context. Data availability at equispaced nodes is a more common situation than availability at node sets that are clustered toward the ends of an integration interval in ways that are specific to particular numerical methods.

7.1 The Trapezoidal Rule and the Newton–Cotes Formulas

The TR uses function values at equispaced nodes:

$$\int_{x_0}^{x_N} f(x)dx = h\sum_{k=0}^{N} f(x_k) - \frac{h}{2}\left[f(x_0) + f(x_N)\right] + O(h^2).$$

For smooth functions, it is very accurate over periodic intervals, but usually quite inaccurate in nonperiodic cases. Historically, the TR has been traced back to Babylonian astronomers around 50 BC (Ossendrijver, 2016). Schematically, its pattern of weights can be illustrated as

$$h\left\{\frac{1}{2} \quad 1 \quad 1 \quad 1 \quad 1 \quad \cdots \quad 1 \quad 1 \quad 1 \quad \frac{1}{2}\right\}.$$

Simpson's formula

$$\frac{h}{3}\{1 \quad 4 \quad 2 \quad 4 \quad 2 \quad \cdots \quad 4 \quad 2 \quad 4 \quad 1\}$$

follows from forming pairs of adjacent subintervals and for each of these integrating a quadratic interpolant.[1] Collecting the subintervals in groups of 3, groups of 4, and so on leads to the Newton–Cotes formulas, described in all standard numerical textbooks. We will not discuss these here, also because the Euler–Maclaurin and Gregory formulas (Sections 7.2 and 7.3) are more conceptually sound in utilizing that the dominant TR errors arise at the interval ends. Corrections to reach higher orders are best done at these ends, rather than compromising the accuracy throughout the interior of the interval.[2]

7.2 The Euler–Maclaurin Formulas

About 50 years after Leibniz and Newton first presented calculus (in 1684 and 1687, respectively), Leonhard Euler[3] and Colin Maclaurin[4] (independently) discovered an asymptotic series of correction terms to the TR:

$$\begin{aligned}\int_{x_0}^{x_N} f(x)dx \sim &\left(h\sum_{k=0}^{N} f(x_k) - \frac{h}{2}\left[f(x_0)+f(x_N)\right]\right)\\ &+\frac{h^2}{12}\left[f^{(1)}(x_0)-f^{(1)}(x_N)\right]-\frac{h^4}{720}\left[f^{(3)}(x_0)-f^{(3)}(x_N)\right]\\ &+\frac{h^6}{30240}\left[f^{(5)}(x_0)-f^{(5)}(x_N)\right]\\ &-\frac{h^8}{1209600}\left[f^{(7)}(x_0)-f^{(7)}(x_N)\right]+-\cdots,\end{aligned} \tag{7.1}$$

with coefficients obtained from the generating function

$$\begin{aligned}\frac{1}{e^z-1}-\frac{1}{z} &= -\frac{1}{2}+\frac{1}{12}z-\frac{1}{720}z^3+\frac{1}{30240}z^5-\frac{1}{1209600}z^7+-\cdots\\ &= \sum_{k=1}^{\infty}\frac{B_k}{k!}z^{k-1}.\end{aligned} \tag{7.2}$$

Here, B_k denotes the Bernoulli numbers, and $h = \frac{x_N - x_0}{N}$.

When using the *midpoint rule* (MR) instead of TR over the same interval

[1] Alternatively obtained after one step of Richardson extrapolation (Section E.1) of TR results. This rule was used by Kepler in 1615, long before being published by Simpson in 1743.

[2] If the TR gives d accurate digits for a smooth periodic function, Simpson's rule gives only about $d/2$ digits. This drops to $d/3$ digits, $d/4$ digits, and so on, for the sequence of Newton–Cotes methods – see Section 7.7.2.1.

[3] 1707–1783, Swiss mathematician, physicist, and astronomer.

[4] 1698–1746, Scottish mathematician.

$[x_0, x_N]$, the nodes are placed at the equispaced interior locations $\xi_k = x_0 + \left(k + \frac{1}{2}\right) h$, with $k = 0, 1, \ldots, N-1$. In this case, the error expansion becomes[5]

$$\begin{aligned}\int_{x_0}^{x_N} f(x)dx \sim &\left(h \sum_{k=0}^{N-1} f(\xi_k)\right) \\ &- \frac{h^2}{24}\left[f^{(1)}(x_0) - f^{(1)}(x_N)\right] + \frac{7\,h^4}{5760}\left[f^{(3)}(x_0) - f^{(3)}(x_N)\right] \\ &- \frac{31\,h^6}{967680}\left[f^{(5)}(x_0) - f^{(5)}(x_N)\right] \\ &+ \frac{127\,h^8}{154828800}\left[f^{(7)}(x_0) - f^{(7)}(x_N)\right] - + \cdots,\end{aligned} \tag{7.3}$$

with coefficients obtained from

$$\frac{e^{z/2}}{e^z - 1} - \frac{1}{z} = -\frac{1}{24}z + \frac{7}{5670}z^3 - \frac{31}{967680}z^5 + - \cdots = -\sum_{k=1}^{\infty} \frac{B_k}{k!}(1 - 2^{1-k})z^{k-1}. \tag{7.4}$$

The original derivations by Euler and Maclaurin are described in Ferraro (2008); Mills (1985). Numerous more recent derivations are available, for example, in Apostol (1999); Arakawa et al. (2014); Atkinson (1988); Ralston and Rabinowitz (1978) as well as three different ones in Appendix A of Fornberg (2020). Two main conclusions are apparent from the EML formulas:

i. *The leading errors in the TR come from the ends of the interval.* The usual second-order accuracy (error $O(h^2)$) becomes fourth order if $f'(x_0) = f'(x_N)$ and sixth order if also $f'''(x_0) = f'''(x_N)$. For periodic problems, the accuracy increases beyond any algebraic order (assuming $f(x)$ is infinitely differentiable). This case is surveyed in Trefethen and Weideman (2014); Weideman (2002).
ii. *It is at the ends that one should adjust quadrature weights if one wants to increase the accuracy of a TR-type scheme.*

The EML formulas give the difference between an integral and a sum. If either the integral or the sum is available, they provide a means to approximate the other. A primary difficulty in applying the EML formulas (and which the methods described later aim to numerically overcome) is that including many terms requires knowledge of correspondingly high-order derivatives. Unless $f(x)$ is a very simple function, algebraic complexity can increase very fast under repeated analytic differentiations.

[5] Often known as "the second Euler–Maclaurin formula."

7.3 Gregory's Method

Already in 1670, James Gregory[6] found an effective TR end correction procedure, using a collection of mathematical tools that are typically attributed to later calculus pioneers. This work also represents a very early use of FD approximations of high orders of accuracy. For simplicity, we first describe it when applied to the difference $\int_0^\infty f(x)dx - \sum_{k=0}^\infty f(k)$. With the notation $\Delta f(k) = f(k+1) - f(k)$ for the first one-sided difference, it follows that

$$\begin{aligned} \Delta^0 f(0) &= f(0) \\ \Delta^1 f(0) &= f(1) - f(0) \\ \Delta^2 f(0) &= f(2) - 2f(1) + f(0) \\ \Delta^3 f(0) &= f(3) - 3f(2) + 3f(1) - f(0) \\ \vdots \quad & \quad \vdots \end{aligned} \tag{7.5}$$

with coefficients from Pascal's triangle. Gregory's idea was to look for an improved TR formula of the form

$$\int_0^\infty f(x)dx \sim \left(\sum_{k=0}^\infty f(k)\right) + \left[b_0\Delta^0 + b_1\Delta^1 + b_2\Delta^2 + \cdots\right] f(0)\,, \tag{7.6}$$

and then to use only a finite number of these correction terms. The exponential test function substitution $f(x) = e^{-zx}$ in (7.6)[7] gives

$$\frac{1}{z} = \frac{1}{1-e^{-z}} + \left[b_0 - b_1(1-e^{-z}) + b_2(1-e^{-z})^2 - b_3(1-e^{-z})^3 + - \cdots\right].$$

With the further substitution

$$w = (1-e^{-z})\,, \tag{7.7}$$

that is, $z = -\log(1-w)$, this becomes

$$\frac{1}{\log(1-w)} + \frac{1}{w} = -b_0 + b_1 w - b_2 w^2 + b_3 w^3 - + \cdots\,. \tag{7.8}$$

The coefficients b_k can now be calculated recursively based on the Taylor expansion of $\log(1-w)$:

$$\begin{aligned} &b_0 = -\frac{1}{2},\ b_1 = \frac{1}{12},\ b_2 = -\frac{1}{24},\ b_3 = \frac{19}{720}, \\ &b_4 = -\frac{3}{160},\ b_5 = \frac{863}{60480},\ b_6 = -\frac{275}{24192},\ldots. \end{aligned} \tag{7.9}$$

[6] 1638–1675, Scottish mathematician and astronomer.

[7] This same substitution in $\int_0^\infty f(x)dx \sim \sum_{k=0}^\infty f(k) + \sum_{n=0}^\infty \alpha_n f^{(n)}(0)$ is one way to arrive at (7.1), (7.2).

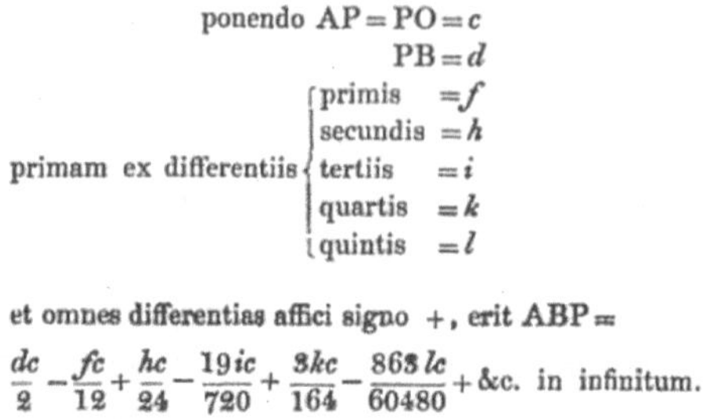

ponendo AP = PO = c
PB = d
primam ex differentiis { primis = f, secundis = h, tertiis = i, quartis = k, quintis = l }

et omnes differentias affici signo +, erit ABP =
$\frac{dc}{2} - \frac{fc}{12} + \frac{hc}{24} - \frac{19ic}{720} + \frac{3kc}{164} - \frac{863lc}{60480} +$ &c. in infinitum.

Figure 7.1 Brief extract from the bottom of page 208 and the top of page 209 in Gregory (1670). We recognize here both (7.6) and (7.9). The coefficient error (164 vs. 160) was a writing mistake in the original letter.

Using only $b_0 = -\frac{1}{2}$ turns (7.6) into the TR. For each further term, the accuracy order increases by 1.[8] In several instances, Gregory used the expansions that Brook Taylor is known for publishing in 1715. Had it not been for Gregory's early death (from a stroke at age 36), the early history of calculus might have been recorded quite differently.

Table 7.1 gives the Gregory *corrections* d_k, to be added to weights all starting off at 1, up through $p = 10$. In case of a finite interval, we add this d_k-sequence from each end (in reversed order at the right end) to obtain actual weights to use. These two sequences can overlap; each one just has to fit into the interval. If the grid spacing is h rather than 1, all weights need to be multiplied by h.

The EML formula (7.1) provides an alternative way to arrive at the Gregory methods: Approximate as many as possible of its derivatives with one-sided FD approximations of the specified stencil width.

Example 7.3.1 Derive the order $p = 6$ line in Table 7.1 from the EML formula.

Solution Since the Gregory formula for $p = 6$ is 5 nodes wide, we can only approximate the leading EML terms (using $h = 1$) up through the third derivative, giving

$$\begin{array}{rlllllll}
-\frac{1}{2} & \{ & 1, & 0, & 0, & 0, & 0 & \} \text{ TR correction} \\
+\frac{1}{12} & \{ & -\frac{25}{12}, & 4, & -3, & \frac{4}{3}, & -\frac{1}{4} & \} \text{ 1st order., line ``order 4'' in Table 1.3} \\
-\frac{1}{720} & \{ & -\frac{5}{2}, & 9, & -12, & 7, & -\frac{3}{2} & \} \text{ Best 5-node approx. for } f^{(3)}(0) \\
= & \{ & -\frac{193}{288}, & \frac{77}{740}, & -\frac{7}{30}, & \frac{73}{720}, & -\frac{3}{160} & \} \text{ Gregory weight corrections.}
\end{array}$$
□

[8] Gregory's original work on this method is described in a letter he wrote on November 23, 1670, preserved in the *Commercium Epistolicum* collection of manuscripts at the Royal Society, and is reproduced in Turnbull (1939). An extract of the letter is shown in Figure 7.1.

Table 7.1 *Corrections d_k to the weights all being 1, according to Gregory's formulas, up through order $p = 10$. The middle entry on the $p = 10$ line shows the first instance of a correction < -1, leading to a negative quadrature weight $w_k = 1 + d_k$.*

$p =$	Gregory corrections d_k to the weights all being 1									
2	$-\frac{1}{2}$									
3	$-\frac{7}{12}$	$\frac{1}{12}$								
4	$-\frac{5}{8}$	$\frac{1}{6}$	$-\frac{1}{24}$							
5	$-\frac{469}{720}$	$\frac{59}{240}$	$-\frac{29}{240}$	$\frac{19}{720}$						
6	$-\frac{193}{288}$	$\frac{77}{240}$	$-\frac{7}{30}$	$\frac{73}{720}$	$-\frac{3}{160}$					
7	$-\frac{41393}{60480}$	$\frac{23719}{60480}$	$-\frac{11371}{30240}$	$\frac{7381}{30240}$	$-\frac{5449}{60480}$	$\frac{863}{60480}$				
8	$-\frac{12023}{17280}$	$\frac{6961}{15120}$	$-\frac{66109}{120960}$	$\frac{33}{70}$	$-\frac{31523}{120960}$	$\frac{1247}{15120}$	$-\frac{275}{24192}$			
9	$-\frac{2558783}{3628800}$	$\frac{1908311}{3628800}$	$-\frac{299587}{403200}$	$\frac{115963}{145152}$	$-\frac{426809}{725760}$	$\frac{112477}{403200}$	$-\frac{278921}{3628800}$	$\frac{33953}{3628800}$		
10	$-\frac{63887}{89600}$	$\frac{427487}{725760}$	$-\frac{3498217}{3628800}$	$\frac{500327}{403200}$	$-\frac{6467}{5670}$	$\frac{2616161}{3628800}$	$-\frac{24019}{80640}$	$\frac{263077}{3628800}$	$-\frac{8183}{1036800}$	
$\vdots$	$\vdots$	$\vdots$	$\vdots$	$\vdots$	$\vdots$	$\vdots$	$\vdots$	$\vdots$	$\vdots$	$\ddots$

As observed in Section 1.2.2, one-sided FD weights grow rapidly with the stencil width, and the small coefficients in the EML expansion cannot fully compensate for this.[9]

Using the function EC_weights given in a fractional derivative context in Section 8.3.2, the line in Table 7.1 corresponding to a given p-value can be obtained numerically in MATLAB by

```
w = -EC_weights(-1,1,0:p-2);     w(1) = w(1)-1;
```

Alternatively, Gregory weights (including corrections at both ends and the TR weights in-between) can be computed by the code shown as Algorithm 7.1.

For the task of approximating integrals using only equispaced function values from within the integration interval, the Gregory approach is remarkably efficient.[10] For analytic functions, the approach extends also to end corrections using function values at grid points in the complex plane surrounding the interval end points (Section 7.5) and to arbitrarily located points (real or complex; Sections 8.3.1 and 8.3.2).

[9] The growth can be reduced as described in Appendix F.2.

[10] Commonly included in older textbooks, but unfortunately omitted in many modern ones (in favor of Simpson's rule and the Newton–Cotes formulas).

Algorithm 7.1 MATLAB code to compute weights for Gregory quadrature.

```
function w = Gregory(n_nodes,h,order)
% This function calculates the Gregory quadrature weights for
% equispaced integration
%    n_nodes   Total number of nodes
%    h         Step size
%    order     Order of accuracy desired, 2,3,4,... (with 2 giving the
%              trap. rule). The value must satisfy 2 <= order <= n_nodes
%     w        Output: The weights to be used for the successive
%              function values
%
% If f is a row vector containing the function values, the integral is
% approximated by the statement   integral = f*w'

% Create the sequence of Gregory coefficients
r = 1./(1:order); gc = toeplitz(r(1:order-1),[r(1),zeros(1,order-2)])...
       \r(2:order)';

% Create the weights vector w and then update it at the two ends of
% the interval

w                          = ones(1,n_nodes)*h;
w_updates                  = sum(h*repmat(gc,1,order-1).*pascal(order-1,1));
w(1:order-1)               = w(1:order-1)-w_updates(1:order-1);
w(n_nodes-order+2:n_nodes) = w(n_nodes-order+2:n_nodes)...
                            -fliplr(w_updates(1:order-1));
end
```

7.4 FD-based Approximation of Infinite Sums When the Integral Is Available

For this approach, we focus on evaluating an equispaced nonoscillatory infinite sum which, for later notational convenience, we will write as $\sum_{k=0}^{\infty} f(k+\frac{1}{2})$. We further assume that

$$F(x) = -\int_x^{\infty} f(t)dt \tag{7.10}$$

is readily available. As an illustrative example, we obtain from the third row in the FD Table 1.1 in the case of $h = \frac{1}{2}$,

$$\begin{aligned} f(k) \approx 2\Big\{ &-\tfrac{1}{60}F(k-\tfrac{3}{2}) + \tfrac{3}{20}F(k-1) - \tfrac{3}{4}F(k-\tfrac{1}{2}) + 0F(k-0) \\ &+ \tfrac{3}{4}F(k+\tfrac{1}{2}) - \tfrac{3}{20}F(k+1) + \tfrac{1}{60}F(k+\tfrac{3}{2})\Big\}. \end{aligned}$$

Summing this relation for $k = \frac{1}{2}, \frac{3}{2}, \frac{5}{2}, \ldots$, the LHS becomes $\sum_{k=0}^{\infty} f(k+\frac{1}{2})$, and the RHS becomes a telescoping sum, with coefficients for F adding up as follows (at locations $-1, -\frac{1}{2}, 0, \frac{1}{2}, \ldots$):

$$
\begin{array}{lcccccccccccl}
2\{ & -\frac{1}{60}, & \frac{3}{20}, & -\frac{3}{4}, & 0, & \frac{3}{4}, & -\frac{3}{20} & \frac{1}{60}, & 0, & 0, & 0, & 0, & \cdots\} \\
+2\{ & & & -\frac{1}{60}, & \frac{3}{20}, & -\frac{3}{4}, & 0, & \frac{3}{4}, & -\frac{3}{20} & \frac{1}{60}, & 0, & 0, & \cdots\} \\
+2\{ & & & & & -\frac{1}{60}, & \frac{3}{20}, & -\frac{3}{4}, & 0, & \frac{3}{4}, & -\frac{3}{20} & \frac{1}{60}, & \cdots\} \\
+2\{ & & & & & & & -\frac{1}{60}, & \frac{3}{20}, & -\frac{3}{4}, & 0, & \frac{3}{4}, & \cdots\} \\
+2\{ & & & & & & & & & -\frac{1}{60}, & \frac{3}{20}, & -\frac{3}{4}, & \cdots\} \\
+2\{ & & & & & & & & & & & -\frac{1}{60}, & \cdots\} + \cdots \\
=\{ & -\frac{1}{30}, & \frac{3}{10}, & -\frac{23}{15}, & \frac{3}{10}, & -\frac{1}{30}, & 0, & 0, & 0, & 0, & 0, & 0, & \cdots\}.
\end{array}
$$

From the bottom line, we thus read off the approximation

$$\sum_{k=0}^{\infty} f\left(k+\frac{1}{2}\right) \approx -\frac{1}{30}F(-1) + \frac{3}{10}F\left(-\frac{1}{2}\right) - \frac{23}{15}F(0) + \frac{3}{10}F\left(\frac{1}{2}\right) - \frac{1}{30}F(1)\,. \tag{7.11}$$

These weights are given as the third row ($\mu = 3$) in Table 7.2. Each row in this table ($\mu = 1, 2, 3, \ldots$) is obtained in the same way, starting with row μ (accuracy order 2μ) in Table 1.1. Several ways to generate these weights are given in Fornberg (2020).[11] The weight in row μ, x-location $x = k/2$, $k = 0, \pm 1, \pm 2, \ldots, \pm(\mu - 1)$ can also be found in closed form:

$$w_{\mu,k} = (-1)^{k+1} \sum_{n=|k|}^{\mu-1} \frac{(n!)^2}{(2n+1)(n+k)!(n-k)!}. \tag{7.12}$$

From this follows that $|w_{\mu,k}| \leq \sum_{n=0}^{\mu-1} \frac{1}{2n+1}$, implying that there is no risk of dangerous numerical cancellations even if using very high μ-values (high-order accurate formulas).

The example below compares exact EML (based on (7.3)) against the FD approach just described for evaluating a well-known infinite sum. This example is highly nontypical for infinite sums in that the function that is summed can readily be analytically differentiated an arbitrarily number of times without any increase in algebraic complexity. The FD method, not needing any analytic differentiations, is more widely applicable (however, both methods need the integral (7.10) to be available).

[11] One of these alternatives is to approximate the increasing order EML derivatives with standard FD approximations, in this case starting from (7.3) and then applying centered FD formulas. While high-order accurate approximations to low-order derivatives are widely used, this derivation relies on FD approximations to very high-order derivatives. Yet one more derivation is given in Pinelis (2018).

Table 7.2 *Weights for $F(x)$ at different x-locations when approximating $\sum_{k=0}^{\infty} f(k+\frac{1}{2})$.*

	Weights for $F(x)$ at x-locations										
μ	$-\frac{5}{2}$	-2	$-\frac{3}{2}$	-1	$-\frac{1}{2}$	0	$\frac{1}{2}$	1	$\frac{3}{2}$	2	$\frac{5}{2}$
1						-1					
2					$\frac{1}{6}$	$-\frac{4}{3}$	$\frac{1}{6}$				
3				$-\frac{1}{30}$	$\frac{3}{10}$	$-\frac{23}{15}$	$\frac{3}{10}$	$-\frac{1}{30}$			
4			$\frac{1}{140}$	$-\frac{8}{105}$	$\frac{57}{140}$	$-\frac{176}{105}$	$\frac{57}{140}$	$-\frac{8}{105}$	$\frac{1}{140}$		
5		$-\frac{1}{630}$	$\frac{5}{252}$	$-\frac{38}{315}$	$\frac{125}{252}$	$-\frac{563}{315}$	$\frac{125}{252}$	$-\frac{38}{315}$	$\frac{5}{252}$	$-\frac{1}{630}$	
6	$\frac{1}{2772}$	$-\frac{2}{385}$	$\frac{25}{693}$	$-\frac{568}{3465}$	$\frac{1585}{2772}$	$-\frac{6508}{3465}$	$\frac{1585}{2772}$	$-\frac{568}{3465}$	$\frac{25}{693}$	$-\frac{2}{385}$	$\frac{1}{2772}$

Example 7.4.1 Approximate the Euler–Mascheroni constant

$$\gamma = \lim_{n\to\infty}\left(\sum_{k=1}^{n}\frac{1}{k} - \log n\right) = 1 + \sum_{k=2}^{\infty}\left(\frac{1}{k} + \log(k-1) - \log(k)\right)$$
$$\approx 0.57721566490153286061 \tag{7.13}$$

by directly summing $\sum_{k=2}^{N-1}$ and then applying the FD method and the EML formula (7.3) to approximate the remainder $\sum_{k=N}^{\infty}$.

Solution In this example,

$$f(x) = \frac{1}{x} + \log\left(1 - \frac{1}{x}\right),$$
$$f^{(k)}(x) = (-1)^{k+1}(k-1)!\left(\frac{1}{(x-1)^k} - \frac{1}{x^k} - \frac{k}{x^{k+1}}\right), \quad k = 1, 2, 3, \ldots,$$
$$F(x) = 2(x-1)\,\mathrm{arccoth}(2x-1) - 1.$$

Figure 7.2 shows how the resulting accuracy varies with μ for the FD method, as described earlier (solid curves) and for the EML approximation (dashed curves). Here, μ is the number of terms used (for EML derivatives up through order $2\mu - 1$) in each of the three different cases of $N = 10, 60, 300$. The FD approach at level μ uses $(2\mu - 1)$ F-evaluations. In close agreement with an estimate given in Fornberg (2020), the FD approach gives for the same μ approximately $2\mu \log_{10}\left(\frac{\pi}{2}\right) \approx 0.39\mu$ fewer correct decimal digits.[12]

[12] This reference also describes a Hermite-type algorithm (using $(\mu - 1)$ values of $f(x)$ and μ of $F(x)$ near the end point) which nearly eliminates this difference.

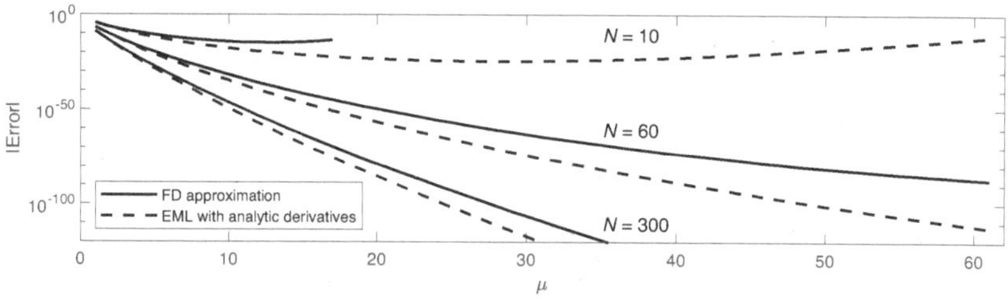

Figure 7.2 Convergence rates for the infinite sum in Example 7.4.1 (having summed $\sum_{k=2}^{N-1}$ explicitly and then accelerated $\sum_{k=N}^{\infty}$).

As μ is increased indefinitely, these asymptotic approximations eventually diverge. However, by summing enough initial terms (increasing N), any desired accuracy level is readily reached. Monitoring the change in result when increasing μ provides practical (but nonrigorous) error estimates. □

7.5 Euler–Maclaurin for Evaluating Integrals Using Centered FD in the Complex Plane

The only information needed about $f(z)$ (analytic) in this approach is that it is numerically available on a 2-D grid in the complex plane, with some spacing h in each direction. Grid-based data is more readily available than values at arbitrary locations in contexts that include:

- Visualization of analytic functions is increasingly often used and is virtually always grid based. FD formulas can repurpose such data for a variety of different tasks with minimal additional processing.
- Some analytic functions are most effectively calculated over gridded domains.[13]

The main present idea (introduced in Fornberg, 2021b) is to first decide on a stencil size (such as $n = 1$; i.e., 3×3), and then replace as many derivatives as possible in (7.1) with their corresponding FD approximations (as shown for some low-order derivatives in Table 6.1). The derivative approximations are combined according to the successive coefficients in the EML formula. We note that the increasing powers of h in the EML formula are perfectly balanced by increasing h-powers in the denominators for increasing derivative orders.

[13] such as the Painlevé transcendents, see Fornberg and Weideman (2011).

With no loss of generality, we can consider $\int_0^\infty f(x)dx$ and focus on the corrections around $x = 0$. Schematically, the regular TR can be summarized as

$$\int_0^\infty f(x)dx = h\left\{\frac{1}{2} \quad 1 \quad 1 \quad 1 \quad 1 \quad 1 \;\ldots\;\ldots\right\} f + O(h^2). \tag{7.14}$$

For example, with 3×3 stencils, odd-order derivatives up through $f^{(7)}(0)$ can be approximated. Added together, we then obtain a 3×3 end corrected TR that can similarly be summarized as

$$\int_0^\infty f(x)dx =$$

$$h\left\{\left[\begin{bmatrix} \boxed{\frac{-821-779i}{403200}} & \boxed{-\frac{1889i}{100800}} & \boxed{\frac{821-779i}{403200}} \\ \boxed{-\frac{1511}{100800}} & \left\{\frac{1}{2}\right. & 1\ \boxed{+\frac{1511}{100800}} \\ \boxed{\frac{-821+779i}{403200}} & \boxed{\frac{1889i}{100800}} & \boxed{\frac{821+779i}{403200}} \end{bmatrix} \quad 1\;1\;1\;1\;\ldots\;\ldots\;\right\}\right\} f + O(h^{10}), \tag{7.15}$$

where the added "correction terms" have been boxed. With a 5×5 stencil, the accuracy becomes $O(h^{26})$. In general, the power is one more than the number of nodes in the stencil. Figure 7.3 illustrates that the correction coefficients are numerically very small compared to the regular weights in the TR. As shown in Fornberg (2023), the extremely high derivatives that enter when stencils are made still larger eventually cause divergence. For size 7×7 (with convergence rate $O(h^{50})$), some correction weights reach around 10 in magnitude.[14]

7.5.1 TR End Correction Weights by the Method of Exponential Test Functions

This method produces identical TR correction stencils more directly than the approach just described and requires no knowledge of the Euler–Maclaurin expansion (neither its coefficients, nor its existence). Following the method of exponential test functions, we form $Le^{z\xi} = \int_0^\infty e^{z\,\xi}dz - h\left(\frac{1}{2} + \sum_{k=1}^\infty e^{kh\xi}\right) = \frac{h}{2}\coth\frac{h\xi}{2} - \frac{1}{\xi}$. The weights w_k at nodes z_k follow then from equating leading powers of ξ between this function and $\sum_{k=1}^n w_k e^{\xi z_k}$. The following Mathematica code (here set to the 3×3 case, and for simplicity using $h = 1$) gives the

[14] This growth in weights with increasing stencil size is of little practical concern (as the accuracy orders are extremely high even for small stencils). Fortunately, it is absent in the method described in Section 7.4 (as in that case, very large stencil sizes may be called for in cases of extreme accuracy requirements).

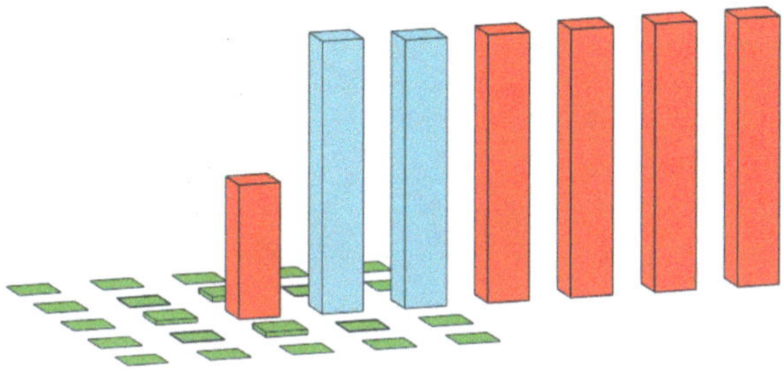

Figure 7.3 Illustration of the magnitudes of the weights in the 5 × 5 case of TR end correction. Entries belonging to the TR formula are shown in red, to the correction stencil in green, and the overlapping entries in blue. Compared to the TR weights (1/2 at the end point(s), and 1 otherwise), the correction stencil weights are two (or more) orders of magnitude smaller. The correction weights in this 5 × 5 stencil case are given in exact rational form in Fornberg (2021b). Figure reproduced from Fornberg (2021b).

correction stencil weights. We use in the code m to specify the stencil size to be $(2m+1) \times (2m+1)$, with then a total of $n = (2m+1)^2$ nodes.

```
m = 1; n = (2m + 1)^2;
zk = Flatten[Table[(i - Ij), {j, -m, m}, {i, -m, m}]];
S = (1/2 Coth[ξ/2] - 1/ξ) - Sum[w[k] E^(zk[[k]] ξ), {k, 1, n}];
M = Solve[LogicalExpand[Series[S, {ξ, 0, n - 1}] == 0]];
MatrixForm[ArrayReshape[Table[w[k]/.M[[1]], {k, 1, n}], {2m + 1, 2m + 1}]]
```

The code produces the output

$$\begin{pmatrix} -\frac{821}{403200}-\frac{779i}{403200} & -\frac{1889i}{100800} & \frac{821}{403200}-\frac{779i}{403200} \\ -\frac{1511}{100800} & 0 & \frac{1511}{100800} \\ -\frac{821}{403200}+\frac{779i}{403200} & \frac{1889i}{100800} & \frac{821}{403200}+\frac{779i}{403200} \end{pmatrix},$$

matching the correction weights shown in (7.15). With the function EC_weights given in Section 8.3.2, this weight matrix is obtained numerically by

```
z = (-m:m)+1i*(m:-1:-m)';   w = EC_weights(-1,1,z);   w(m+1,m+1) = 0;
```

7.5.2 Application to Contour Integration

A main application of the technique in Section 7.5.1 is contour integration of an analytic function between any two grid points or around closed contours. We note in (7.14) and (7.15) that h enters only as a multiplicative factor for all

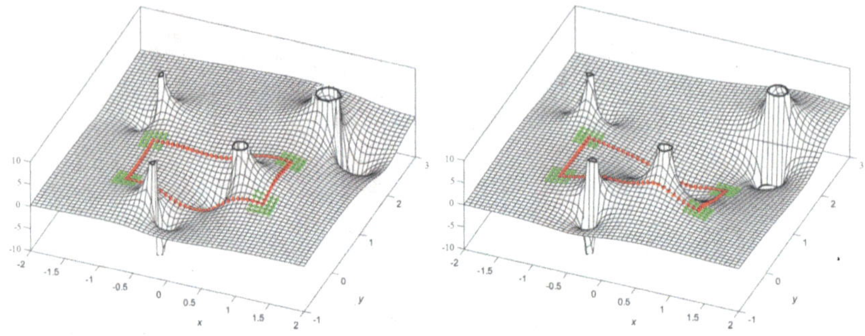

(a) Real part (b) Imaginary part

Figure 7.4 Real and imaginary parts of the test function (7.16), with the integration contour shown by the red dots.

weights. At the end point of an integration along the real axis, the correction stencil should be multiplied by $-h$; similarly by $\pm ih$ for the ends of a vertical line segment. The following example was previously considered in Fornberg (2021b,d):

Example 7.5.1 Integrate

$$f(z) = \frac{2}{z - 0.4(1+i)} - \frac{1}{z + 0.4(1+i)} + \frac{1}{z + (1.2 - 1.6i)} - \frac{3}{z - (1.3 + 2i)} \tag{7.16}$$

around the rectangle indicated by grid points marked by red dots in Figure 7.4.

Solution The green dots indicate the additional function values used when end-correcting each line segment by a 5×5 stencil, as illustrated in Figure 7.3. Since the contour goes around one pole (at $z = 0.4(1 + i)$, with residue 2), the analytic result is $4\pi i$. Figure 7.5 shows the rates at which the errors go to zero as h decreases. The errors originating from the path corners are, in the 5×5 case, so low that, for relatively large h (to the left in the figure), the total error becomes instead dominated by the TR along the interior of the line segments. Also, these errors can be further improved on, as described in Fornberg (2021d). □

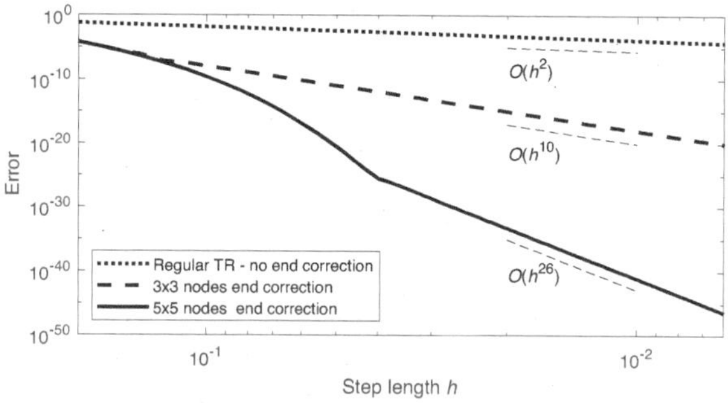

Figure 7.5 Log–log plot of the error for the test problem described in Section 7.5.2, computed in extended precision arithmetic. With the grid spacing $h \approx 0.07$ shown in Figure 7.4, the errors for the three approximations are roughly 10^{-2}, 10^{-9}, and 10^{-15}, respectively. Roughly speaking, the grid density needed for a "reasonably resolved" functional display is comparable to what is needed for double precision integration accuracy when using TR with 5×5 size correction stencils.

7.6 Evaluating Alternating Sums with Terms of Decreasing Magnitude

An "obvious" approach is to group the terms in pairs, to obtain a positive series for which one of the methods described earlier might be applicable. This approach requires that the corresponding integral can be readily evaluated, which is far from always the case. Remaining options fall in two main categories – linear and nonlinear methods. Two examples of the former are given next, in Section 7.6.1, followed by some brief additional comments.

7.6.1 Two Formulas by Euler

The counterpart to the EML formula for an alternating series is known as the Euler–Boole formula (Borwein et al., 2009):

$$\sum_{k=N}^{\infty}(-1)^k f(k) = \frac{(-1)^k}{2}\left\{f(N) - \frac{1}{2}f'(N) + \frac{1}{24}f^{(3)}(N) - \frac{1}{120}f^{(5)}(N) \right.$$
$$\left. + \frac{17}{40370}f^{(7)}(N) - + \cdots\right\}. \tag{7.17}$$

Two observations:

i. There is in this case no integral present – only the value and derivatives of the function f at location N.

ii. There is no step size h to choose/optimize (as the alternating sign pattern in some sense imposes a step size of 1).

This second issue limits the computational utility of (7.17), as this precludes us from utilizing the fact that the successive increasing powers of h in the EML formulas perfectly balanced the powers of h in the denominators of the respective derivative approximations.

The well-known Euler transformation

$$\sum_{k=N}^{\infty}(-1)^k a_k \approx \sum_{k=N}^{\infty}(-1)^k \frac{\Delta^k a_N}{2^{k+1}} \tag{7.18}$$

(with Δ defined as in (7.5)) can be seen as applying one-sided FD approximations to the derivatives in (7.17). Instead using centered FD approximations offers little advantage in this situation.

7.6.2 Extrapolation Approaches

The arguably most successful computational methods for alternating series (such as repeated Aitken extrapolation and Padé conversion) are nonlinear, and not naturally cast in terms of higher-order FD approximations. For these, see Sections E.2.2 and E.3 (Examples E.2.1 and E.3.2), and for comprehensive overviews Brezinski and Zaglia (1991); Nowak (2019); Sidi (2003).

7.6.3 The Abel–Plana Formulas

Before leaving the topic of infinite sums, we note the Abel–Plana formulas (from 1820, named after Niels Henrik Abel[15] and Giovanni Amadeo Plana[16]). These require the terms in the infinite sums to come from a function $f(z)$ that is analytic in the right complex half-plane.[17]

$$\sum_{k=0}^{\infty} f(k) = \int_0^{\infty} f(x)dx \quad + \frac{1}{2}f(0) + i\int_0^{\infty} \frac{f(it)-f(-it)}{e^{2\pi t}-1}dt,$$

$$\sum_{k=0}^{\infty}(-1)^k f(k) = \quad \frac{1}{2}f(0) + i\int_0^{\infty} \frac{f(it)-f(-it)}{2\sinh(\pi t)}dt.$$

Because of the rapid growth in the denominators ($e^{2\pi t} - 1$ and $2\sinh(\pi t)$, respectively), these formulas can sometimes (when applicable) be numerically

15 1802–1829, Norwegian mathematician; best known for the first complete proof that polynomial equations of degree above four cannot in general be solved in radicals.

16 1781–1864, Italian astronomer and mathematician; performed detailed studies of the moon's orbit.

17 with some additional "fine print" required.

attractive. One way to arrive at both (7.1) and (7.17) comes from Taylor expanding the respective integrands around $t = 0$ (however thereby turning exact formulas into asymptotic ones – divergent if their number of terms is increased indefinitely).[18]

7.7 Some Notes on Orders of Convergence

7.7.1 FD Approximations of a Derivative

A Taylor expansion is very accurate in the immediate vicinity of a single point, making it a convenient tool both for deriving FD approximations to a derivative and for determining such approximations' order of accuracy. Section 1.2.1.2 describes an alternative means (exponential test functions) both for finding FD weights and providing accuracy orders. The derivation of the Padé algorithm (Section 1.2.1.5), as sketched out in Example 1.2.5, uses exponential test functions. Both approaches (Taylor expansions and exponential test functions) are well suited for deriving and analyzing FD approximations for derivatives.

7.7.2 Approximations of Integrals

7.7.2.1 Taylor Expansions

Since integrals are not local operators, it is less obvious how to measure orders of accuracy for quadrature methods. Finding the highest degree polynomial for which such a method is exact is a questionable approach for determining convergence rates (as functions of step size h). High-degree interpolating polynomials on equispaced grids feature the Runge phenomenon and are ill suited to represent general functions. For Gregory quadrature, this approach might suggest the incorrect result that accuracy orders increase in steps of 2 (as for the sequence of Newton–Cotes methods) – see the analysis of Gregory errors in Martensen (1964).

One reason for the relative popularity of the Newton–Cotes formulas in recent numerical textbooks may be that Taylor expansion-based error estimates are readily obtained. For example, when integrating over an interval $[a, b]$,

$$\begin{aligned}
\text{TR error} &= -\tfrac{(b-a)}{12}\, h^2 f''(\eta),\\
\text{Simpson error} &= -\tfrac{(b-a)}{180}\, h^4 f^{(4)}(\eta),
\end{aligned}$$

[18] For (7.1), see for example Fornberg and Piret (2020), Section 12.5.2.

where in both cases η is some point within $[a, b]$.[19] Although rigorous, these estimates can be misleading:

i. they do not reflect the spectral convergence rates of these methods when applied to sufficiently smooth periodic functions;
ii. the dominant errors on finite intervals usually arise from end effects;
iii. if applied to convergent integrals on semiinfinite intervals, the difference between integral and (infinite) sum will again be $O(h^2)$ and $O(h^4)$, respectively (in spite of then $b - a = \infty$).

In periodic cases,[20] increasing the order of Newton–Cotes methods is likely to cause a loss rather than a gain in accuracy. The TR method is then exact for all Fourier modes up to the highest that can be represented on a grid with spacing h, that is, up to the modes $e^{\pm i\pi x/h}$. This range is halved for Simpson's rule (since its weights repeat with period 2 rather than period 1), corresponding to a loss of about 1/2 of the significant digits compared to TR (with losses progressively more severe if Newton–Cotes orders are increased).

7.7.2.2 Exponential Test Functions

This approach provides approximations for the Euler–Maclaurin and Gregory quadrature methods in just a few lines of algebra.[21]

It also readily (but less rigorously) provides the order of accuracy for the standard equispaced quadrature schemes, for example,

$$\int_0^\infty e^{\xi x}dx - \left\{\begin{matrix}\text{TR}\\ \text{step } h\end{matrix}\right\} = -\frac{1}{\xi} + \frac{1}{2}h\coth\left(\frac{h\xi}{2}\right) = \frac{\xi}{12}\,h^2 + O(h^4),$$

$$\int_0^\infty e^{\xi x}dx - \left\{\begin{matrix}\text{Simpson}\\ \text{step } h\end{matrix}\right\} = -\frac{1}{\xi} + \frac{1}{3}h(2+\cosh(h\xi))\operatorname{csch}(h\xi) = \frac{\xi^3}{180}\,h^4 + O(h^6).$$

The exponential test function approach formed the basis for the methods in Chapters 6 and 7, and will do so again for computing fractional derivatives (Chapter 8) and enhanced Gregory quadrature (Section F.2). It gives accuracy orders that match numerically observed ones.[22]

[19] Error estimates involving only lower order derivatives are also possible, for example, as given for Simpson's formula in Dragomir et al. (2000).
[20] and similarly, within interval interiors rather than near their ends.
[21] Apply $f(x) = e^{\xi x}$ with Re $\xi < 0$ to $\int_0^\infty f(x)\,dx$ and to an approximation of desired form, and then choose parameters to eliminate as many powers of h as possible.
[22] For Gregory quadrature, see for example Figure 8.4(a). The top curve corresponds to $p = 2$, and displays $O(h^2)$ convergence, with the power of h increasing by one for each successive case.

8
Fractional-Order Derivatives

The idea of derivatives of noninteger orders is almost as old as calculus and, with that, integer order derivatives. In 1695, Guillaume de l'Hôpital[1] asked Leibniz about derivatives of order 1/2, to which Leibniz responded with some analysis and a comment along the lines of "This is an apparent paradox from which, one day, useful consequences will be drawn." In 1823, Abel presented an extensive framework for fractional-order calculus, as well as a first physical application.

While integer-order derivatives are uniquely defined, numerous different definitions have been proposed for derivatives of noninteger (including complex) orders, with different features and areas of applicability.[2] Noninteger-order derivatives are often defined as inverses of certain fractional-order integrals, and as such are not local operators,[3] but have two key points – a *base point* and an *evaluation point* and they depend on an integral between them. For a differential equation with a fractional derivative in time, this integral allows past states to influence solutions in additional ways to what is possible to model with integer-order time derivatives (which take into account only the current state). We discuss numerical approximations to fractional derivatives here since the FD-based quadrature tools that were described in Chapter 7 can often be applied very effectively.

There is extensive literature available on many aspects (mathematical, application oriented, etc.) of fractional derivatives, see for example Diethelm

[1] 1661–1704, French mathematician and aristocrat, best known for l'Hôpital's rule for evaluating $0/0$ and ∞/∞ limits (first published by l'Hôpital but discovered and forwarded to him by Johann Bernoulli).

[2] Many seemingly different fractional derivative versions are, in Valério et al. (2022), shown to belong to a two-parameter family. Complex orders were considered already by Liouville (1832).

[3] Local fractional-order derivatives (with somewhat different properties) have also been considered Abdeljawad (2015); Guzmán et al. (2018); Khalil et al. (2014).

(2010); Gorenflo and Mainardi (2008); Kilbas et al. (2006); Podlubny (1990); Samko et al. (1993). A brief summary is given in Ishteva (2005). The description below partly follows Fornberg and Piret (2023); Fornberg et al. (2025), and is focused on the numerical evaluation of fractional derivatives using the quadrature techniques described in Chapter 7. This present chapter concludes with some comments on fractional powers of the Laplace operator.

8.1 Riemann–Liouville and Caputo Derivatives

The two most commonly used definitions are:
Riemann–Liouville,

$$ {}_{a}^{\mathrm{RL}}D_t^{\alpha} f(t) = \frac{1}{\Gamma(n-\alpha)} \frac{d^n}{dt^n} \int_a^t \frac{f(\tau)}{(t-\tau)^{\alpha+1-n}} d\tau \,, \qquad n-1 < \alpha < n, \quad (8.1) $$

and Caputo (1967),

$$ {}_{a}^{\mathrm{C}}D_t^{\alpha} f(t) = \frac{1}{\Gamma(n-\alpha)} \int_a^t \frac{\frac{d^n}{d\tau^n} f(\tau)}{(t-\tau)^{\alpha+1-n}} d\tau \,, \qquad n-1 < \alpha < n. \quad (8.2) $$

With no loss of generality, we let the *base point* a be located at $a = 0$ and (initially) assume that the *evaluation point* t satisfies $t > 0$ (later to be generalized to an arbitrary placed point in the complex plane). The two definitions are closely related,

$$ {}_{0}^{\mathrm{RL}}D_t^{\alpha} f(t) = {}_{0}^{\mathrm{C}}D_t^{\alpha} f(t) + \sum_{k=0}^{n-1} \frac{t^{k-\alpha}}{\Gamma(k+1-\alpha)} f^{(k)}(0), \quad (8.3) $$

but have nevertheless quite different character, as illustrated in Figure 8.1 in the cases of differentiating $f(t) = e^t$ for $0 \le \alpha \le 1$, $0 \le t \le 1$.[4] Figure 8.2 similarly displays the Caputo derivative of $\sin t$ for $-2.5 \le \alpha \le 5.5, 0 \le t \le 2$.[5] Our focus on Caputo rather than on Riemann–Liouville derivatives in the rest of this chapter is largely motivated by the simplicity they offer when formulating initial conditions for fractional order differential equations, as described in Podlubny (1990, Chapter 4), and Ishteva (2005, Chapters 2, 4)[6]

We simplify the notation in the following by writing ${}_{0}^{\mathrm{C}}D_t^{\alpha} f(t)$ as $D^{\alpha} f(t)$.

[4] The closed form values in this parameter range are ${}_{0}^{\mathrm{RL}}D_t^{\alpha} e^t = e^t \left(1 - \frac{\Gamma(-\alpha,t)}{\Gamma(-\alpha)}\right)$ and ${}_{0}^{\mathrm{C}}D_t^{\alpha} e^t = e^t \left(1 - \frac{\Gamma(1-\alpha,t)}{\Gamma(1-\alpha)}\right)$.

[5] Since $\left.\frac{d^k}{dt^k} \sin t\right|_{t=0} = 0$ for $k = 0, 2, 4, \ldots$, the Caputo derivative is discontinuous with respect to α only at $\alpha = 1, 3, 5, \ldots$ rather than at all positive integers.

[6] The differential equation ${}_{0}^{\mathrm{C}}D_t^{\alpha} y(t) = f(t, y(t))$ requires $\lceil \alpha \rceil$ initial conditions; $y(0)$ for $0 < \alpha \le 1$, $y(0)$ and $y'(0)$ for $1 < \alpha \le 2$, and so on.

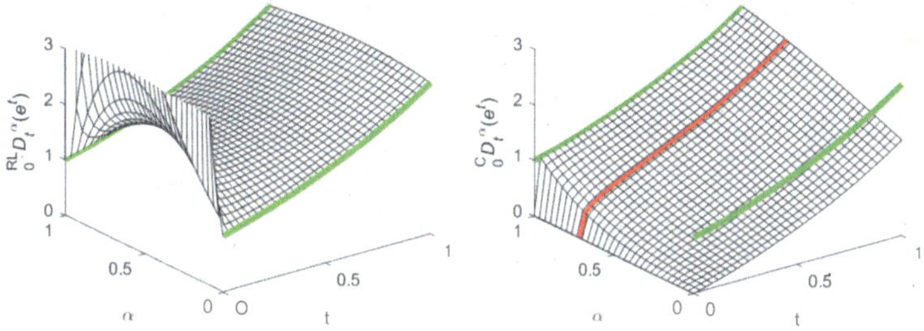

(a) Riemann–Liouville derivative (b) Caputo derivative

Figure 8.1 Riemann–Liouville and Caputo derivatives of order $\alpha \in [0, 1]$ for $f(t) = e^t$, $t \geq 0$. The green curves show the standard derivatives for the integer-order cases $\alpha = 0$ and $\alpha = 1$. The red curve in part (b) corresponds to $\alpha = \frac{5}{7}$, and is seen again in Figure 8.5.

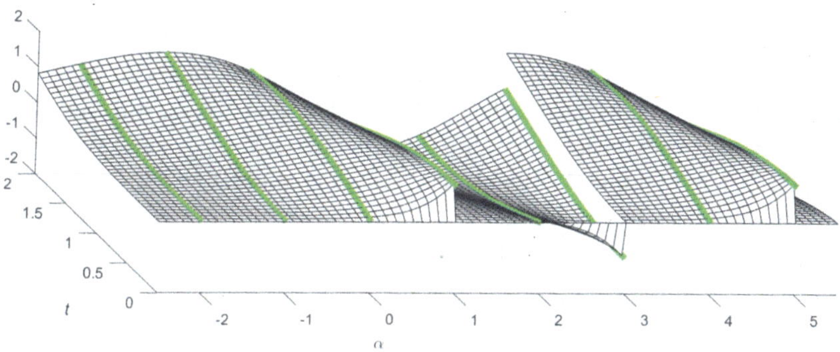

Figure 8.2 The Caputo derivative ${}^{C}_{0}D_t^{\alpha} \sin t$ for $-2.5 \leq \alpha \leq 5.5, 0 \leq t \leq 2$. The green curves show the values in cases of α integer. This example is atypical since for most functions, there will be discontinuities also for $\alpha = 2, 4, 6, \ldots$.

Since, for m integer, $D^{\alpha+m} f(t) = D^{\alpha} D^{m} f(t)$,[7] we furthermore focus on $n = 1$, that is, $0 < \alpha < 1$.[8]

[7] In the Riemann–Liouville case, the operators D^{α} and D^{m} need to be in reversed order for this result. In both cases, the operators commute if $f^{(k)}(0) = 0$ for $k = 0, 1, 2, \ldots, m$.

[8] Another convenience with the Caputo version is that (8.2) implies that $D^{\alpha}\{\text{constant}\} = 0$. On the other hand (as seen in Figure 8.1), while $\lim_{\alpha \to 1} D^{\alpha} f(t) = f'(t)$, $\lim_{\alpha \to 0} D^{\alpha} f(t) = f(t) - f(0)$. In the Riemann–Liouville case, both limits are as expected.

8.2 Introductory Comments on Some Approximation Methods

The Grünwald–Letnikov sum (introduced in 1868)

$$\Sigma_{\text{GL}} = \frac{1}{h^{\alpha}} \sum_{j=0}^{[t/h]} (-1)^j \binom{\alpha}{j} f(t - jh) \tag{8.4}$$

satisfies

$$\lim_{h \to 0} \Sigma_{\text{GL}} =_{a}^{\text{RL}} D^{\alpha} f(t). \tag{8.5}$$

However, its convergence rate is only $O(h^1)$, severely limiting its computational utility.

Chapter 2 of Li and Zeng (2015) contains an extensive survey of numerical methods for fractional derivatives. Several of the described approaches produce orders of accuracy $p = 2$ or $p = 3$ (including a second-order accurate Grünwald–Letnikov enhancement). However, the authors note (on page 71) "Although higher-order schemes can be constructed as above, we still think that it is necessary to reconsider the pth-order schemes ($p = 2, 3, \ldots, 6$) with new methods." The examples of time-stepping fractional order ODEs (in Section 3.3.1 of this reference) have either first- or second-order convergence. Approximations to Caputo derivatives of accuracy orders up to $4 - \alpha$ are described in Mokhtari and Mostajeran (2020); Ramezani et al. (2020).

For functions that are well represented as expansions in orthogonal (e.g., Jacobi) polynomials,[9] certain identities are available for fractional derivative (and integral) calculations, as utilized for example in Chebfun.[10]

8.3 Gregory-based Approximations on an Equispaced Grid along the Real Axis

Following the FD theme of this book, we focus this Section 8.3 and Section 8.4 on the cases when the $f(t)$-function is smooth, and with values available only at equispaced grid points, along the real axis, and in the complex plane (the latter described further in Fornberg and Piret (2023)).

[9] Generally, not the case for fractional differential equations, as these typically feature base point singularities.

[10] Derived from equation (18.17.9) in Olver et al. (2010), see also Li and Zeng (2015), Section 2.1.7.

8.3.1 Approximations Based on TR with End Corrections

We focus on the Caputo case and generalize the Gregory method from Section 7.3. For simplicity, the description below assumes the derivative order $0 < \alpha < 1$. The task is then to approximate

$$D^{\alpha} f(t) = \frac{1}{\Gamma(1-\alpha)} \int_0^t \frac{f'(\tau)}{(t-\tau)^{\alpha}} d\tau \,, \tag{8.6}$$

where (following our FD theme) we assume $f(\tau)$ (rather than $f'(\tau)$) to be available numerically, and only at h-spaced τ-values. Integration by parts leads us to consider[11]

$$\alpha \int_0^t \frac{f(\tau)}{(t-\tau)^{\alpha+1}} d\tau \tag{8.7}$$

and the TR-type sum

$$\alpha h \sum_{k=1}^{[\frac{t}{h}]-1} \frac{f(kh)}{(t-kh)^{\alpha+1}} \tag{8.8}$$

corresponding to this integral over the interior of the interval $[0, t]$ (in contrast to regular TR, omitting a term at each end).[12]

Lower end: There is no singularity present in (8.7) around $\tau = 0$, and the TR end corrections can be applied similarly to descriptions in Chapter 7 and Section F.2. The weights when end correcting at $\tau = 0$ can be obtained by the method of exponential test functions, considering $\int_0^{\infty} g(\tau) d\tau$ with $g(\tau) = e^{\xi\,\tau}$, with Re $\xi < 0$, in which case

$$\begin{aligned} &\int_0^{\infty} e^{\xi\,\tau} d\tau - h \sum_{k=1}^{\infty} e^{\xi\,kh} \\ &= \frac{h}{2}\left(1 + \coth \frac{h\xi}{2}\right) - \frac{1}{\xi} \\ &= -h\left(\frac{\zeta(0)}{0!} + \frac{\zeta(-1)}{1!}(h\xi)^1 + \frac{\zeta(-2)}{2!}(h\xi)^2 + \frac{\zeta(-3)}{2!}(h\xi)^3 + \cdots\right). \end{aligned} \tag{8.9}$$

In this expansion, the terms with $\xi^2, \xi^4, \xi^6, \ldots$ vanish (corresponding to the trivial zeros of the zeta function). For arbitrary locations of nodes in end correction stencils, the weights can be obtained from the coefficients in this Taylor expansion as described in Section 8.3.2.[13]

[11] Temporarily omitting the factor $\frac{1}{\Gamma(1-\alpha)}$ in (8.6). Note also that this integral, as it stands, diverges at $\tau = t$. This divergence cancels out in the described end correction procedure.

[12] to avoid a divide by zero at $k = \left[\frac{t}{h}\right]$.

[13] At this lower end, we need to remember that the integration by parts leading to (8.7), also produced a term $-\frac{f(0)}{t^{\alpha}}$.

Upper end: This end correction can be obtained by comparing (8.8) against $\int_0^t \frac{f'(\tau)}{(t-\tau)^\alpha}d\tau$ (without integrating by parts). To simplify the notation, we analyze also this upper end correction around the origin, and thus change variables $\sigma = t - \tau$, and set $g(\sigma) = f(t-\tau)$. This leads us to consider the end correction problem (around $\sigma = 0$) of the form

$$\int_0^\infty \frac{g'(\sigma)}{\sigma^\alpha}d\sigma - \alpha h \sum_{k=1}^{\infty} \frac{g(kh)}{(kh)^{\alpha+1}}\,. \tag{8.10}$$

The exponential test function substitution becomes now $g(\sigma) = e^{\xi\,\sigma}$. Assuming as before $\operatorname{Re}\xi < 0$, we obtain

$$\begin{aligned}&\int_0^\infty \frac{\xi\, e^{\xi\,\sigma}}{\sigma^\alpha}d\sigma - \alpha h \sum_{k=1}^{\infty} \frac{e^{\xi kh}}{(kh)^{\alpha+1}}\\ &= (-\xi)^{\alpha-1}\xi\,\Gamma(1-\alpha) - \alpha h^{-\alpha}\mathrm{Li}_{1+\alpha}(e^{h\xi})\\ &= -\alpha h^{-\alpha}\left(\frac{\zeta(\alpha+1)}{0!} + \frac{\zeta(\alpha)(h\xi)}{1!} + \frac{\zeta(\alpha-1)(h\xi)^2}{2!} + \frac{\zeta(\alpha-2)(h\xi)^3}{3!} + \cdots\right),\end{aligned} \tag{8.11}$$

where Li denotes the polylogarithm function (also known as Jonquière's function). The two terms in the first RHS of (8.11) each contain a fractional power of ξ that cancels out in their difference. Just as for the nonsingular lower end (8.9), the result in (8.11) is again a Taylor series in ξ.

Combined procedure: The TR-type part of the calculation (8.8) is corrected separately at the lower and upper ends. At both ends, we have obtained error expressions in the form of Taylor expansions, which we will convert to correction weights. At the lower end, these weights are applied to values of $\frac{f(\tau)}{(t-\tau)^{\alpha+1}}$ (with an additional term $-\frac{f(0)}{t^\alpha}$ that has arisen from the integration by parts). At the upper end, the correction weights are applied instead to $f(\tau)$-values.[14] The resulting approximation can be written as

$$\int_0^t \frac{f'(\tau)}{(t-\tau)^\alpha}d\tau = \underbrace{-\alpha h \sum_{k=1}^{[t/h]-1} \frac{f(kh)}{(t-kh)^{\alpha+1}}}_{\text{TR-type sum}} + \underbrace{\left(-\frac{f(0)}{t^\alpha} + \alpha\, h \sum_{k=1}^{n} w_k^{(\mathrm{bp})} \frac{f(z_k^{(\mathrm{bp})})}{(t-z_k^{(\mathrm{bp})})^{\alpha+1}}\right)}_{\text{Lower end correction}}$$

[14] Here with no integration by parts term, as the integral in (8.10) is in its original form.

$$+\underbrace{\left(\alpha h^{-\alpha}\sum_{k=1}^{n} w_k^{(\mathrm{ep})} f(t - z_k^{(\mathrm{ep})})\right)}_{\text{Upper end correction}} + O(h^n). \tag{8.12}$$

Here, n denotes the total number of stencil nodes at each end (and is not related to derivative order, as in (8.1), (8,2)), $z_k^{(\mathrm{bp})}$ and $z_k^{(\mathrm{ep})}$ are the locations of the correction stencil nodes relative to the base and evaluation points,[15] and $w_k^{(\mathrm{bp})}$ and $w_k^{(\mathrm{ep})}$ the respective weights, obtained as described in the next Section 8.3.2. At both ends, the node locations for the end corrections can be chosen arbitrarily.[16]

8.3.2 Conversion of Taylor Expansions in ξ to Weights in a Correction Stencil

With correction stencil nodes located at z_k, $k = 1, \ldots, n$ (grid-based or not), equating a leading part of a Taylor expansion $\sum_{k=0}^{n-1} \alpha_k \xi^k$ to that of a correction stencil $\sum_{k=1}^{n} w_k e^{\xi z_k}$ leads to the linear system

$$\begin{bmatrix} 1 & 1 & \cdots & \cdots & 1 \\ z_1 & z_2 & \cdots & \cdots & z_n \\ z_1^2 & z_2^2 & \cdots & \cdots & z_n^2 \\ \vdots & \vdots & & & \vdots \\ z_1^{n-1} & z_2^{n-1} & \cdots & \cdots & z_n^{n-1} \end{bmatrix} \begin{bmatrix} w_1 \\ w_2 \\ w_3 \\ \vdots \\ w_n \end{bmatrix} = \begin{bmatrix} 0!\alpha_0 \\ 1!\alpha_1 \\ 2!\alpha_2 \\ \vdots \\ (n-1)!\alpha_{n-1} \end{bmatrix}. \tag{8.13}$$

The conversion of the Taylor expansions in the last large bracket in (8.9) and (8.11) to weights at node locations z_k can be done by calling the MATLAB function in Algorithm 8.1 for $\alpha = -1$ and $0 < \alpha < 1$, respectively.[17]

8.3.3 Numerical Illustrations of Gregory-based Calculations

Figure 8.3(a) shows a test function that is infinitely differentiable over $[0, 1]$, and part (b) its analytic Caputo derivative of order $\alpha = 1/2$ $\Big($with a square root type singularity at the origin: $D^{1/2} f(t) = -\frac{400}{\sqrt{\pi}} t^{1/2}\left(1 - \frac{400}{9} t + \frac{6400}{9} t^2 + \cdots\right)\Big)$. Figure 8.4(a) shows the error as function of the number of Gregory correction coefficients used at each end (1–13, from top and down, corresponding to

[15] We denote these by $z_k^{(\mathrm{bp})}$, $z_k^{(\mathrm{ep})}$ rather than $\tau_k^{(\mathrm{bp})}$, $\tau_k^{(\mathrm{ep})}$ as they may be placed in the complex plane surrounding the base and evaluation points.

[16] As for Gregory quadrature, inward from the two end points, or surrounding these along the real axis or in the complex plane case.

[17] For the case of real-valued z_k-values, an $O(n^2)$ solution method for Vandermonde linear systems is described in Björck and Pereyra (1970); Demmel and Koev (2005).

Algorithm 8.1 Algorithm for converting Taylor coefficients to weights.

```
function w = EC_weights(al,h,z)
% Calculate weights in a stencil matching the Taylor expansion
% ζ(α+1)/0!+ζ(α)(hξ)/1!+ζ(α−1)(hξ)^2/2!+ζ(α−2)(hξ)^3/3!+...
% Input parameters
%    al  The parameter denoted α above
%     h  Step size in TR-type approximation (with first term omitted)
%     z  Stencil node locations (array or vector)
% Output parameter
%     w  Weights at locations given by z

dm = size(z); z = z(:); sq = (0:length(z)-1);     % Make z a column vector
A = rot90(vander(z)); rhs = zeta(al+1-sq).*h.^sq;
     % Form vandermonde coefficient matrix and RHS
w = A\rhs'; w = reshape(w,dm);
     % Solve linear system and arrange weights as for z
end
```

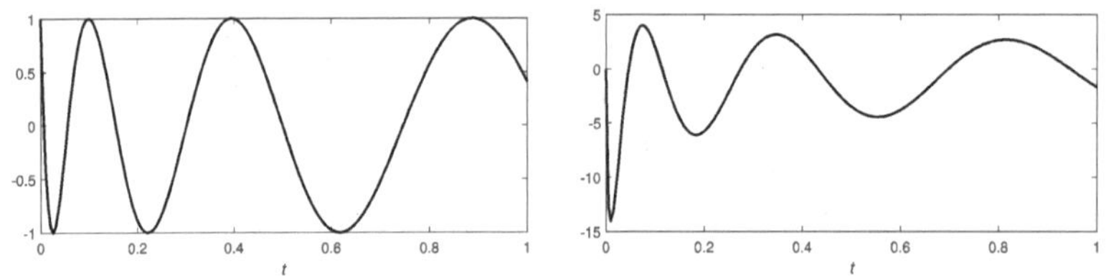

(a) $f(t) = \cos(20\sqrt{t})$ (b) $D^{1/2} f(t) = -10\sqrt{\pi}\mathrm{H}_0(20\sqrt{t})$

Figure 8.3 (a) The test function used in Section 8.3.3 (and later in Sections F.2 and G.2.2); (b) Its Caputo fractional derivative. Here H_0 denotes the Struve function of order zero (alternatively expressed as a hypergeometric function: $\mathrm{H}_0(t) = \frac{2t}{\pi}\, {}_1F_2\left(1; \frac{3}{2}, \frac{3}{2}; -\frac{t^2}{4}\right)$).

accuracy orders 2–14, with the first case being the trapezoidal rule TR). Part (b) shows for the same number of correction terms at each end the error when using Gregory-type approximations to the integral (8.6), implemented as described Section 8.3.1. Unsurprisingly, as derivative calculations in general are less stable than integrations,[18] the rounding error level for small h is notably higher.[19] The slope of each curve in the Caputo derivative case is a factor of $h^{-3/2}$ less than in the regular integral case. Nevertheless, rates of convergence better than $O(h^{10})$ are readily achieved.[20]

[18] While all Gregory weights are positive for accuracy orders up through $p = 9$, this is never the case for Caputo derivative approximations, since $D^{\alpha}(1) = 0$, implying that the weights must add up to zero.

[19] Here using standard double precision.

[20] The quintic spline-based method described in Ciesielski (2024) (requiring quad precision arithmetic) converges at the rate $O(h^{6-\alpha})$.

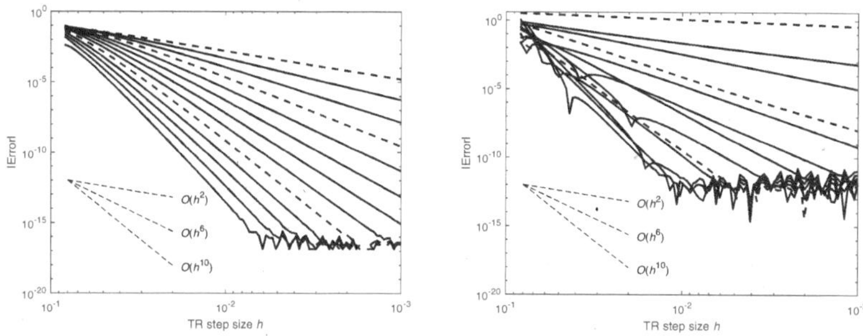

(a) Gregory integration (b) Gregory-based Caputo differentiation

Figure 8.4 Errors as functions of the number of Gregory-type correction coefficients applied at each interval end (1–13 for the curves, from top to bottom) and functions of step size h, when applied to the test function shown in Figure 8.3(a). In the cases of using 1, 4, 9 correction coefficients, the curves are shown dashed instead of solid.

8.4 Fractional Derivative Approximations for an Analytic Function Given on a Grid in the Complex Plane

We first note that, with $f(\tau)$ an analytic function, the fractional derivative

$$D^{\alpha} f(z) = \frac{1}{\Gamma(1-\alpha)} \int_0^z \frac{f'(\tau)}{(z-\tau)^{\alpha}} d\tau, \quad 0 < \alpha < 1,$$

becomes again an analytic function.

Denoting $D^{\alpha} f(z)$ by $g(z)$, it follows from integration by parts that $g'(z) = \frac{1}{\Gamma(1-\alpha)} \left(\frac{f'(0)}{z^{\alpha}} + \int_0^z \frac{f''(\tau)}{(z-\tau)^{\alpha}} d\tau \right)$. With both $g(z)$ and $g'(z)$ well defined, analyticity of $D^{\alpha} f(z) = g(z)$ follows.

8.4.1 Numerical Approach for Complex-valued Evaluation Point

We assume again that $f(z)$ is nonsingular at the base point $z = 0$.[21] The main idea is to follow horizontal and vertical grid lines from the base point (B) to the evaluation point (E), as described for contour integrals in Section 7.5.2 (i.e., with correction stencils also at each path corner). We typically use correction stencils of size 5×5, with weights for the line segment ending in the evaluation point obtained as described in Sections 8.3.1 and 8.3.2. A key requirement is that no correction stencil, centered either at B or at a path corner point, can be close to E (since the integrand has a singularity there, and these correction stencils would then become inaccurate). Corner points close to E are readily

[21] For the case when $f(z)$ is singular at $z = 0$, see Fornberg and Piret (2023).

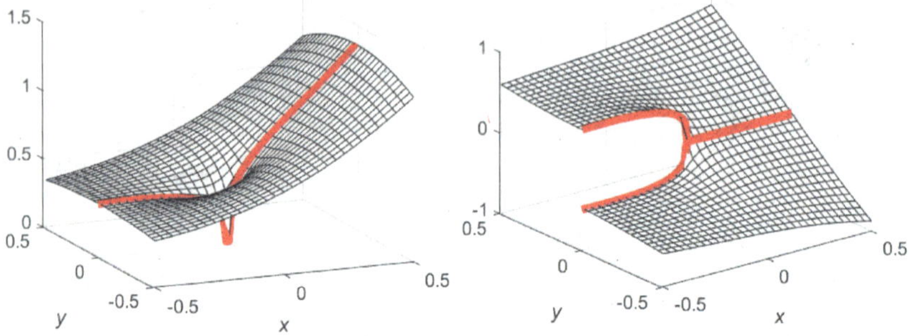

(a) Real part (b) Imaginary part

Figure 8.5 Plots of the real and imaginary parts of $D^{\alpha}e^{z}$ with $\alpha = 5/7$, displayed and computed with $h = 0.04$. The exact value is $D^{\alpha}e^{z} = e^{z}\left(1 - \frac{\Gamma(1-\alpha,z)}{\Gamma(1-\alpha)}\right)$. The red curve to the right of the origin for $\mathrm{Re}(D^{\alpha}f)$ matches the one in Figure 8.1(b).

avoided by suitable path choices, while the case of B and E close can be handled by a somewhat different (and equally accurate) approach also described in more detail in Fornberg and Piret (2023).

8.4.2 Some Illustrations of Fractional Derivatives of Analytic Functions

Figures 8.5 and 8.6 show two cases of fractional derivatives computed using the approach outlined earlier. Stencils of size 5×5 readily suffice for achieving machine precision 10^{-16} across the regions displayed. For more illustrations, see Fornberg and Piret (2023); Higgins (2022).

8.5 Fractional Laplace Operator

The Laplace operator,

$$\Delta = \left(\frac{\partial^2}{\partial x_1^2} + \frac{\partial^2}{\partial x_2^2} + \cdots + \frac{\partial^2}{\partial x_n^2}\right), \tag{8.14}$$

and Laplace's equation $\Delta u = 0$ are ubiquitous in applied mathematics.[22] Fractional powers of it, especially $(-\Delta)^{\alpha}$ with $0 < \alpha < 1$, arise in a range of applications, and can be numerically implemented in relatively straightforward ways. For reviews both on applications and on computational methods for fractional Laplacians, see Daoud and Laamri (2022); D'Elia et al. (2020); Garofalo

[22] It is in the fractional derivative literature quite common to denote the LHS of (8.14) by Δ^2 rather than, as here, by Δ.

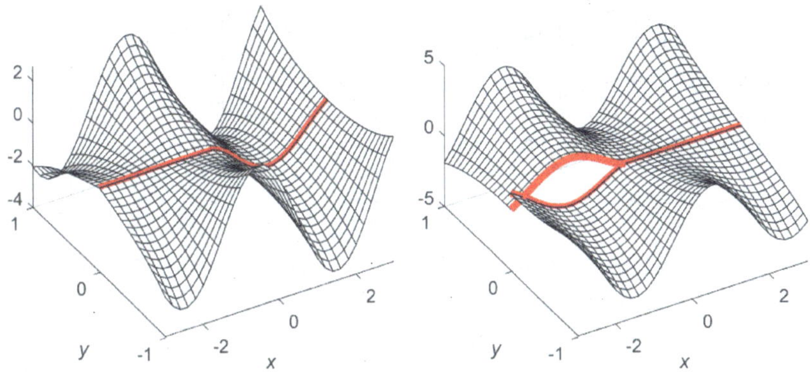

(a) Real part (b) Imaginary part

Figure 8.6 Plots of the real and imaginary parts of $D^\alpha \cos\left(\frac{\pi}{2}z\right)$ with $\alpha = 1/2$, displayed and computed with $h = 0.1$. The exact value is $D^{1/2}\cos\left(\frac{\pi}{2}z\right) = \sqrt{\pi}\left(\cos\left(\frac{\pi}{2}z\right)S(\sqrt{z}) - \sin\left(\frac{\pi}{2}z\right)C(\sqrt{z})\right)$, where $S(z)$ and $C(z)$ are the Fresnel sine and cosine functions.

(2019); Hofreither (2020); Lischke et al. (2020); Zayernouri et al. (2024); Zhuang et al. (2024) (which contain many further references).[23] As in the case for fractional derivatives, the fractional Laplacian is a nonlocal operator, causing issues when applying it on bounded domains as well as when attempting regular (local) FD approximations. We focus most of the comments below on infinite or periodic domains in 3-D.

Two primary application areas of fractional Laplace operators are:

1. modeling anomalous diffusion (and its mathematical equivalents); and
2. solving Cauchy and continuation problems for Laplace's equation.

In the former case, the dimension of the Laplace operator typically matches that of the physical space, while it is one less in the latter case.

8.5.1 Cauchy and Continuation Problems for the 3-D Laplace Equation

We write the 3-D Laplace equation (in x, y, z-space) as

$$\frac{\partial^2 u}{\partial z^2} = -\left(\frac{\partial^2}{\partial x^2} + \frac{\partial^2}{\partial y^2}\right)u = (-\Delta)^1 u \tag{8.15}$$

[23] Fractional Helmholtz equations are studied in Glusa et al. (2021).

(with Δ now denoting the 2-D Laplacian). The Cauchy problem[24] for (8.15), with data given on the plane $z = 0$, is notoriously ill posed. However, one is often interested only in solutions that decay in one direction (such as for z decreasing). The Dirichlet-to-Neumann problem amounts to finding $\frac{\partial u}{\partial z}\big|_{z=0}$ from values of $u|_{z=0}$. This leads us to consider

$$\frac{\partial u}{\partial z} = (-\Delta)^{1/2} u. \tag{8.16}$$

With an appropriate interpretation of $(-\Delta)^{1/2}$ and applied twice, this becomes (8.15) (now well posed in the negative z-direction). Three numerically different but mathematically equivalent options for computing $(-\Delta)^{\alpha} u$ are described next. To be specific, we focus on (8.16).[25] Generalizations to $\alpha \neq \frac{1}{2}$ and to different number of dimensions follow the same patterns.

In 2-D, both the Cauchy and continuation problems for Laplace's equation $\left(\frac{\partial^2}{\partial x^2} + \frac{\partial^2}{\partial y^2}\right) u = 0$ can be addressed by the Hilbert transform (e.g., Fornberg and Piret, 2020, Section 9.4).

8.5.2 Fourier Transform

In the x, y-plane $z = 0$, the Fourier transform (FT) can be defined as

$$\hat{u}(\omega_x, \omega_y) = \frac{1}{(2\pi)^2} \int_{-\infty}^{\infty} \int_{-\infty}^{\infty} u(x, y)\, e^{-i\,(x\,\omega_x + y\,\omega_y)}\, dx\, dy \tag{8.17}$$

and inverted by

$$u(x, y) = \int_{-\infty}^{\infty} \int_{-\infty}^{\infty} \hat{u}(\omega_x, \omega_y)\, e^{i\,(x\,\omega_x + y\,\omega_y)}\, d\omega_x\, d\omega_y \tag{8.18}$$

(see the 1-D FT in Table C.1). Continuing u according to (8.16) from $z = 0$ to $z \neq 0$ is achieved by changing $\hat{u}(\omega_x, \omega_y)$ to $e^{z\sqrt{\omega_x^2+\omega_y^2}}\,\hat{u}(\omega_x, \omega_y)$ before the inversion step.[26] For the Dirichlet-to-Neumann problem, one similarly multiplies $\hat{u}(\omega_x, \omega_y)$ by $\sqrt{\omega_x^2 + \omega_y^2}$. Numerical approximations to the 2-D FTs, (8.17), (8.18), are readily provided by the FFT algorithm, as described in Appendix C.

8.5.3 Principal Value Integral

The Fourier procedure in Section 8.5.2 can be interpreted as saying that $(-\Delta)^{1/2} u$ is a convolution between u and a singular kernel function, thereby

[24] Here, initial-value problems with given values on a line in 2-D or plane in 3-D.

[25] That is, with Δ applied in 2-D.

[26] The fact that $\sqrt{\omega_x^2 + \omega_y^2}$ has a discontinuous derivative at the origin in the ω_x, ω_y-plane tells that the fractional Laplace operator (in contrast to the regular one) cannot be spatially local in the x, y-plane.

leading to a principal value (PV) integral. In the more compact notation $\underline{x} = (x, y, \ldots)$, the general result in d dimensions becomes

$$(-\Delta)^{\alpha} u = \frac{\alpha\, 4^{\alpha}\Gamma(\alpha + \frac{d}{2})}{\pi^{d/2}\,\Gamma(1-\alpha)}\ \mathrm{PV}\int_{R^d} \frac{u(\underline{x}) - u(\underline{x}')}{|\underline{x} - \underline{x}'|^{d+2\alpha}}\, d\underline{x}'. \tag{8.19}$$

Convolutions are often best handled numerically as products in Fourier space. However, this may not be an option on bounded domains. Several $O(h^2)$-accuracy schemes have been proposed for approximating (8.19), see Lanzara et al. (2022) for a higher-order scheme (and for further references).

8.5.4 Radial Basis Functions

Data $u(x, y, 0)$ given on the plane $z = 0$ (or on a surface in $\underline{x} = (x, y, z)$-space) can be approximated by a linear combination of RBFs:

$$u \approx \sum_{k=1}^{N} \lambda_k \phi(||\underline{x} - \underline{x}_k||) \tag{8.20}$$

(cf. Section 5.3). Gaussian radial functions $\phi(r) = e^{-(\varepsilon r)^2}$ are often attractive, as they are relatively local (unless ε is very small). We focus on this case below. For each such $\phi(r)$, we can solve the continuation (Cauchy) problem since, with $r = \sqrt{x^2 + y^2}$,

$$f(r, z) = \frac{1}{2}\int_0^{\infty} \xi\, e^{-\xi^2/4} e^{\varepsilon\,\xi\, z} J_0(\varepsilon\,\xi\, r)\, d\xi \tag{8.21}$$

satisfies[27]

$$\begin{aligned} &f(r, 0) = e^{-(\varepsilon r)^2}, \\ &f(r, z) \longrightarrow 0 \text{ for } z \longrightarrow -\infty, \\ &\Delta f(r, z) = \tfrac{\partial^2 f}{\partial r^2} + \tfrac{1}{r}\tfrac{\partial f}{\partial r} + \tfrac{\partial^2 f}{\partial z^2} \equiv 0. \end{aligned} \tag{8.22}$$

When carrying out RBF-based continuation with (8.21) for some fixed distance $|z|$, $f(r, z)$ needs to be evaluated for a very large number of r-values. As Figure 8.7(a) shows, $f(r, z)$ is a very smooth function in both of its variables. A good computational strategy is to, for the z-value of interest, create a 1-D interpolant that can be rapidly evaluated.

From (8.21) it follows that

$$\left.\frac{\partial^k f(r, z)}{\partial z^k}\right|_{z=0} = (2\varepsilon)^k \Gamma\left(1 + \frac{k}{2}\right)\, {}_1F_1\left(1 + \frac{k}{2}; 1; -(\varepsilon r)^2\right) \tag{8.23}$$

[27] with the third equation equivalent to Laplace's equation in 3-D cylindrical coordinates.

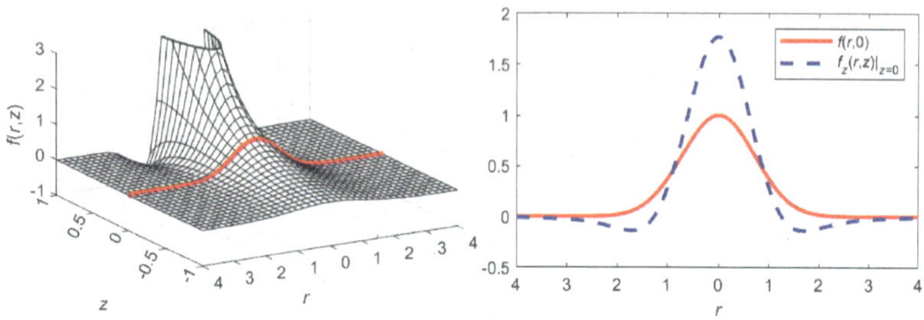

(a) $f(r,z)$ (b) $f(r,0)=e^{-r^2}$ and $\left.\frac{\partial f(r,z)}{\partial z}\right|_{z=0}$

Figure 8.7 The functions in equations (8.21) and (8.25) in the case of $\varepsilon = 1$.

$$= (2\varepsilon)^k \Gamma\left(1+\frac{k}{2}\right)\, e^{-(\varepsilon r)^2} L_{k/2}((\varepsilon r)^2), \tag{8.24}$$

$k = 0, 1, 2, \ldots$. In (8.21)–(8.24), J denotes the J-Bessel function, ${}_1F_1$ the $p = q = 1$ hypergeometric function, and $L_{k/2}$ the degree-$k/2$ Laguerre polynomial (generalized if k is odd; then expressible in a combination of I_0 and I_1 Bessel functions). In particular, we obtain[28]

$$\left.\frac{\partial f(r,z)}{\partial z}\right|_{z=0} = \varepsilon\sqrt{\pi}\, e^{-\frac{1}{2}(\varepsilon r)^2}\left((1-(\varepsilon r)^2)I_0\left(\tfrac{1}{2}(\varepsilon r)^2\right) + (\varepsilon r)^2 I_1\left(\tfrac{1}{2}(\varepsilon r)^2\right)\right), \tag{8.25}$$

$$\left.\frac{\partial^2 f(r,z)}{\partial z^2}\right|_{z=0} = 4\varepsilon^2 e^{-(\varepsilon r)^2}\left(1-(\varepsilon r)^2\right), \tag{8.26}$$

$$\left.\frac{\partial^3 f(r,z)}{\partial z^3}\right|_{z=0} = 2\varepsilon^3\sqrt{\pi}\, e^{-\frac{1}{2}(\varepsilon r)^2}\Big((3-6(\varepsilon r)^2+2(\varepsilon r)^4)I_0\left(\tfrac{1}{2}(\varepsilon r)^2\right) + 2(\varepsilon r)^2(2-(\varepsilon r)^2)I_1\left(\tfrac{1}{2}(\varepsilon r)^2\right)\Big), \tag{8.27}$$

$$\left.\frac{\partial^4 f(r,z)}{\partial z^4}\right|_{z=0} = 16\varepsilon^4 e^{-(\varepsilon r)^2}\left(2-4(\varepsilon r)^2+(\varepsilon r)^4\right). \tag{8.28}$$

Equation (8.25) can be used to solve the Dirichlet-to-Neumann problem for the 3-D Laplace equation when given data expressed as a combination of Gaussian RBFs. Relations similar to some of those above have been given previously also for other radial functions and in arbitrary number of dimensions (Burkardt et al., 2021; Zhuang et al., 2024). Figure 8.7 illustrates the functions $f(r,z)$ and $\left.\frac{\partial f(r,z)}{\partial z}\right|_{z=0}$ as given in (8.21) and (8.25), respectively.

[28] The even-order derivative cases follow also immediately from the last equation in (8.22).

Appendix A
Polynomial Interpolation

With nodes x_k (that are distinct) and function values $f(x_k) = y_k$, $k = 1, 2, \ldots, N$, there is a unique interpolation polynomial $p_{N-1}(x)$ of degree $N-1$. The Lagrange formula will show that it exists. If there were two possibilities, $p_{N-1}(x)$ and $q_{N-1}(x)$, their difference $p_{N-1}(x) - q_{N-1}(x)$ would be an $N-1$-degree polynomial with N zeros, that is, the difference would have to be identically zero, and $p_{N-1}(x) \equiv q_{N-1}(x)$.[1]

A.1 Lagrange's Interpolation Formula

With nodes x_k, $k = 1, 2, \ldots, N$, the polynomials

$$L_j(x) = \frac{\prod_{k=1, k \neq j}^{N} (x - x_k)}{\prod_{k=1, k \neq j}^{N} (x_j - x_k)} \tag{A.1}$$

are of degree $N-1$ and satisfy

$$L_j(x_k) = \begin{cases} 1, & j = k \\ 0, & \text{otherwise} \end{cases} \qquad j, k = 1, 2, \ldots, N.$$

Therefore,

$$p_{N-1}(x) = \sum_{j=1}^{N} f(x_j)\, L_j(x) \tag{A.2}$$

satisfies both requirements (degree $N-1$ and interpolating at N points). Figure 1.3 shows the Lagrange polynomials $L_j(x)$ in case of nine unit-spaced nodes over $[0, 8]$. For larger numbers of nodes, the problem with spurious oscillations grows increasingly severe, as discussed in Section 1.4.[2]

[1] This result does not generalize beyond 1-D, see Section 5.1.

[2] The interpolation formula (A.1), (A.2) was discovered twice (by Waring in 1776 and Euler in 1783) before being described by Lagrange in 1795.

A.2 Some Alternative Polynomial Interpolation Approaches

Although $p_{N-1}(x)$ is unique, there are numerous different algorithms available to arrive at this polynomial, and then to represent it. Notable versions include:

i. **Newton's interpolation formula.** Like Lagrange's interpolation formula, the Newton version is described in most introductory numerical textbooks. It generalizes to Hermite-type interpolation when additional derivative information is available at some nodes. A potential disadvantage is that the node ordering can affect numerical stability.

ii. **Barycentric interpolation.** The polynomial interpolant is in this case written as

$$p_{N-1}(x) = \ell(x) \sum_{j=1}^{N} \frac{w_j}{x - x_j} f(x_j) = \frac{\sum_{j=1}^{N} \frac{w_j}{x-x_j} f(x_j)}{\sum_{j=1}^{N} \frac{w_j}{x-x_j}}, \tag{A.3}$$

where

$$\ell(x) = \prod_{j=1}^{N} (x - x_j), \quad w_j = \frac{1}{\ell'(x_j)} = \frac{1}{\prod_{k \neq j}(x_j - x_k)}, \quad j = 1, 2, \ldots, N. \tag{A.4}$$

With the values for w_j given in (A.4), multiplying both the numerator and the denominator in the rightmost expression in (A.3) by $\prod_{k=1}^{N}(x-x_k)$ makes the numerator into Lagrange's interpolation formula for the data $f(x_j)$, and the denominator becomes identically 1 (in the form of a Lagrange interpolation polynomial for data identical to 1). Strengths include good computational speed and numerical stability. Barycentric interpolation is described, for example, in Rutishauser (1976) and Berrut and Trefethen (2004) (with the latter authors arguing that barycentric interpolation deserves to be known as the standard method of polynomial interpolation).[3] Generalizations to barycentric rational approximations are described in Floater and Hormann (2007); Hofreither (2021); see also Nakatsukasa et al. (2018).

iii. **FD weights algorithm.** If node locations x_j and the interpolation point x do not change, but the function values $f(x_j)$ do, the algorithm in Section 1.2.1.4 creates weights (in its case of approximating the zeroth derivative) which can rapidly be reapplied to each new set of function values. This algorithm generalizes immediately from interpolation to derivative evaluations. If furthermore $f'(x_j)$ values are provided, the algorithm in Fornberg (2021a) provides weights also for these.

[3] Additional error analysis is given in Higham (2004). Some ideas can be traced back to A. F. Möbius ("Der barycentrische Calcul," 1827).

Table A.1 *Costs for repeated interpolations.*

	$O(N^2)$ cost when changing	$O(N)$ cost when changing
Newton	$\{x_j\}$ or $\{f(x_j)\}$	x
Barycentric	$\{x_j\}$	x or $\{f(x_j)\}$
FD weights	$\{x_j\}$ or x	$\{f(x_j)\}$

iv. **Direct solution of linear system.** With $p_{N-1}(x) = a_1+a_2x+\cdots+a_Nx^{N-1}$, the polynomial coefficients follow from solving

$$\begin{bmatrix} 1 & x_1 & \cdots & x_1^{N-1} \\ 1 & x_2 & \cdots & x_2^{N-1} \\ \vdots & \vdots & & \vdots \\ 1 & x_N & \cdots & x_N^{N-1} \end{bmatrix} \begin{bmatrix} a_1 \\ a_2 \\ \vdots \\ a_N \end{bmatrix} = \begin{bmatrix} f(x_1) \\ f(x_2) \\ \vdots \\ f(x_N) \end{bmatrix}. \tag{A.5}$$

While this approach is simple to implement, it is usually not competitive due to cost and conditioning. It offers, however, another demonstration (to the one in Section 1.2.1.1) that a Vandermonde matrix (for distinct nodes) is nonsingular.[4]

Table A.1 compares the operation count between the methods i–iii when the input values are changed for x (interpolation point), $\{x_j\}$ (node points) and $\{f(x_j)\}$ (function values at node points).

A.3 Polynomial Interpolation Error

The most often quoted formula for the error in polynomial interpolation is

$$f(x) - p_{N-1}(x) = \frac{f^{(N)}(\xi)}{N!} \prod_{j=1}^{N} (x - x_j). \tag{A.6}$$

Here, ξ is some point between the smallest and largest of $x_1, x_2, \ldots, x_N, x$. While this formula separates the influences of the node distribution and of the function that is interpolated, both the unknown ξ and the presence of the Nth derivative limits its practicality. If $f(z)$ is an analytic function,[5] a theoretically

[4] Combine the results: (i) For any square matrix A, $\det(A) = \det(A^T)$; (ii) As shown earlier, the interpolation problem has a unique solution; and (iii) If a linear system $A\underline{x} = \underline{b}$ has a unique solution, A is square and nonsingular.

[5] Thus, writing z in place of x.

important alternative is given by[6]

$$f(z) - p_{N-1}(z) = \frac{\ell(z)}{2\pi i} \int_C \frac{f(t)}{\ell(t)(t-z)}\,dt. \tag{A.7}$$

Here $\ell(z) = \prod_{j=1}^{N}(z - z_j)$ and the contour C encloses (in the positive direction) the node points z_j and the interpolation point z, but no singularity of $f(z)$. It is discussed in some detail in Trefethen (2013, Chapter 11); see also Mayers (1966) and Fornberg (1996, Appendix E).

[6] Commonly attributed to Hermite in 1878, but stated earlier by Cauchy in 1826.

Appendix B
Splines

A spline is a piecewise polynomial function, commonly used to interpolate or approximate one-dimensional data. At its nodes, it changes from one polynomial to another, in such a way that a certain number of derivatives remain continuous.[1]

B.1 Introduction

Splines were introduced in Schoenberg (1946). Later generalizations extend to interpolating both gridded and unstructured data in any number of dimensions. General reference books on splines include de Boor (2001); Dierckx (1993); Powell (1981); Prautzsch et al. (2002); Schumaker (1981).[2]

Splines have numerous attractive features as a robust general-purpose approach for interpolating (and thereby also differentiating and integrating) data.[3] Figure B.1(a) shows the lowest-degree polynomial interpolant to somewhat random data, and (b) the piecewise linear interpolant to it. Both versions have undesirable features, with either excessive oscillations or "kinks" (discontinuous first derivative). Denoting the interpolating function $s(x)$, both of these features make the integral $\int s''(x)^2 dx$ over the interval (a rough measure of total amount of curvature) unnecessarily large. Natural cubic splines (c) turn out to minimize this integral over all possible interpolating functions, ensuring both a good level of smoothness together with absence of the Runge phenomenon.

[1] Typically, one derivative less than the degree of the polynomial pieces.
[2] This appendix abbreviates the description in Fornberg (2021c).
[3] In particular, in cases when the data points are not equispaced.

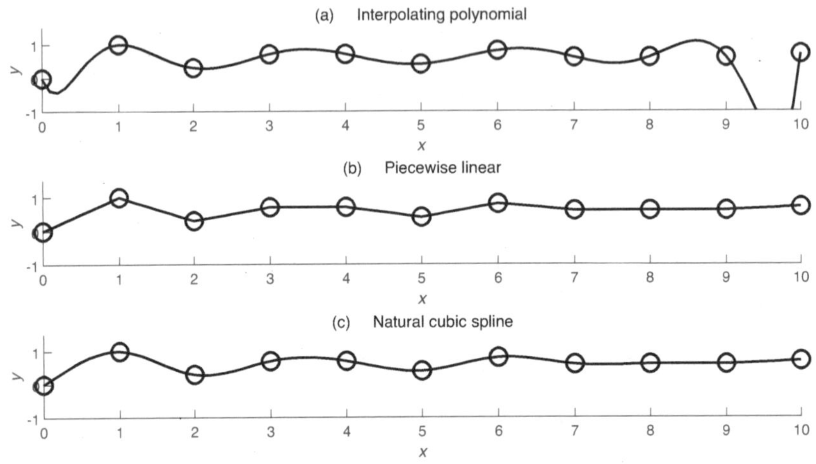

Figure B.1 Contrasting polynomial interpolation with the piecewise linear and the (natural) cubic spline when interpolating the same equispaced data.

B.2 Some Spline Features

For practical use of splines, most coding environments contain a rich set of convenient library routines (including for differentiation, integration, etc.) We therefore focus here on their main concepts, sometimes in simplified cases.

B.2.1 Spline End Conditions

Using between nodes polynomials of degree p (each requiring $p+1$ coefficients), imposing the function value at each node, and continuity of $p-1$ derivatives at each interior node are readily seen to leave us short of $p-1$ pieces of information. We thus need no extra information in the linear ($p = 1$) case (as to be expected for a piecewise linear interpolant), but we are two conditions short in the cubic ($p = 3$) case. Unless specifying otherwise, we focus in the following on this cubic case. Four common options for end conditions are:

1. Set $s'' = 0$ at both end points. This gives the *natural cubic spline*. While of theoretical interest (e.g., minimization of $\int s''(x)^2 dx$, as noted earlier), the information $s'' = 0$ is typically outright wrong. If nodes are spaced h apart, the error when interpolating a smooth function with a cubic spline scales in the interval interior as $O(h^4)$, but the incorrect information $s'' = 0$ reduces this to $O(h^3)$ near the ends.

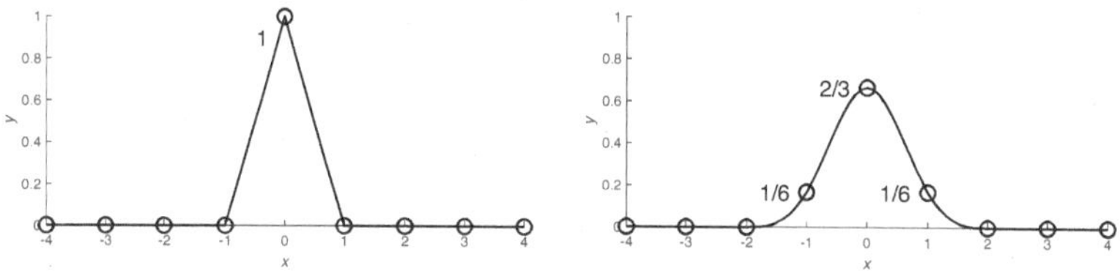

(a) Linear B-spline (b) Cubic B-spline

Figure B.2 Examples of B-splines on unit-spaced grids, with the conventional normalization of integrating to one. Figure reproduced from Fornberg (2021c).

2. Instead of adding two conditions, one can get the counting right also by taking away two freedoms, such as disallowing a discontinuity in s''' one step in from each boundary – known as *Not-a-Knot.*
3. Match the slope at each end point to that of the cubic polynomial that interpolates the four points nearest to each end.[4]
4. If one happens to know s' at each end point, that information can be imposed, giving a *clamped spline*.

Increasing the order of accuracy p requires additional boundary information.

B.2.2 Cubic Splines Used for Derivative Approximations

On a grid with spacing h, the first derivative of a cubic spline is remarkably accurate at the node points since, at all interior ones x_k, it satisfies exactly the relation

$$\frac{1}{6}s'(x_k-h)+\frac{2}{3}s'(x_k)+\frac{1}{6}s'(x_k+h) = \frac{1}{h}\left(-\frac{1}{2}s(x_k - h) + \frac{1}{2}s(x_k + h)\right). \quad \text{(B.1)}$$

This is identical to a fourth-order accurate compact FD approximation (obtained for example by setting $s = 0$, $d = 2$, $n = 2$, $m = 1$ in the algorithm in Section 1.2.1.5).[5] However, the second derivative of an equispaced cubic spline is only accurate to $O(h^2)$ (in domain interiors), that is, no better than the simple explicit approximation in Example 1.2.2. The third derivative of a cubic spline is piecewise constant between nodes and is not uniquely defined at the nodes.

[4] The default choice in MATLAB.

[5] In the interval $[x_k - h, x_k + h]$, the cubic spline $s(x)$ is a linear combination of a cubic and a multiple of $|x - x_k|^3$ (with the latter providing the jump in the third derivative). Equation (B.1) is exact for all cubics as well as for all functions that are even around x_k.

B.2.3 *B*-splines on Equispaced Nodes

It is natural to seek a "simplest" spline, and then represent a general spline as a linear combination of translates of this. A good such choice is the spline that transitions from identical zero nontrivially back to identical zero in the narrowest way possible; see Figure B.2.[6] We will mention just three *B*-spline relations here (in case of unit-spaced nodes and centered at the origin):

1. With $B_p(x)$ denoting the *B*-spline of order p, and

$$B_0(x) = \begin{cases} 1 & -\frac{1}{2} < x < \frac{1}{2}, \\ 0 & \text{otherwise.} \end{cases}$$

   $B_p(x)$ (for $p = 1, 2, 3, \ldots$) is the convolution of $B_{p-1}(x)$ with $B_0(x)$.
2. The Fourier transform of a *B*-spline is remarkably simple:

$$\int_{-\infty}^{\infty} B_p(x) \cos \omega x \, dx = \left(\frac{\sin(\omega/2)}{\omega/2} \right)^{p+2}.$$

3. Derivatives (and integrals) of *B*-splines can be conveniently evaluated by the relation $B'_p(x) = B_{p-1}(x + \frac{1}{2}) - B_{p-1}(x - \frac{1}{2})$.

B.2.4 Equispaced Splines as Combinations of *B*-splines

Figure B.3 shows schematically (as circles) some data $y(i)$, $i = 0, 1, \ldots, N$. Any cubic spline $s(x)$ over these nodes can be written as a linear combination of translates of the cubic *B*-spline: $s(x) = \sum_{i=-1}^{N+1} \alpha_i B_i(x)$ (the subscript on B now marks the location at which the *B*-spline is centered – which includes one extra node outside each interval end). The first task is to determine the α_i-coefficients, as the spline then can be readily evaluated at any location. In the linear system (B.2), the top and bottom equations come from the extra conditions that have to be imposed (shown here for the natural spline), while the remaining equations enforce that the spline takes the correct values at all nodes:

$$\begin{bmatrix} 1 & -2 & 1 & & & \\ \hline 1 & 4 & 1 & & & \\ & 1 & 4 & 1 & & \\ & & \ddots & \ddots & \ddots & \\ & & & 1 & 4 & 1 \\ \hline & & & 1 & -2 & 1 \end{bmatrix} \begin{bmatrix} \alpha_{-1} \\ \alpha_0 \\ \alpha_1 \\ \vdots \\ \alpha_N \\ \alpha_{N+1} \end{bmatrix} = 6 \begin{bmatrix} 0 \\ y_0 \\ y_1 \\ \vdots \\ y_N \\ 0 \end{bmatrix} \begin{matrix} \leftarrow \text{Left BC} \\ \leftarrow \text{Match} \\ \leftarrow \text{values at} \\ \leftarrow \text{nodes} \\ \vdots \\ \leftarrow \text{Right BC} \end{matrix} \tag{B.2}$$

[6] The customary normalization is that the integral evaluates to 1.

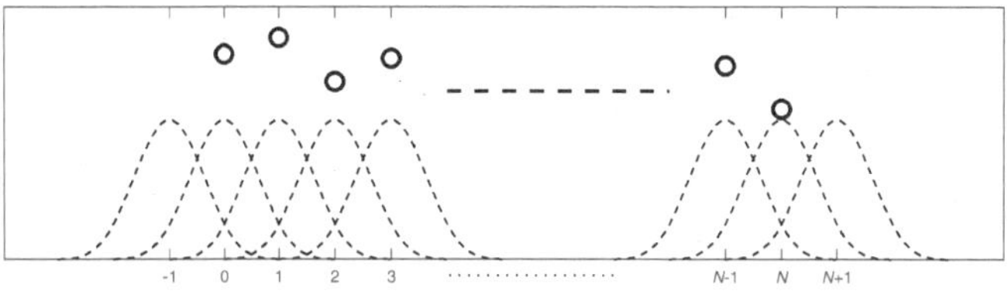

Figure B.3 Schematic illustration of data at $x = 0, 1, 2, \ldots, N$ and B-splines centered at $x = -1, 0, 1, 2, \ldots, N, N+1$. Figure reproduced from Fornberg (2021c).

The 3-group of entries in the top row (and similarly in the bottom row) are obtained from the fact that

$$\left[\begin{array}{ccc} B''_{-1}(0) & B''_0(0) & B''_1(0) \end{array}\right] = \left[\begin{array}{ccc} 1 & -2 & 1 \end{array}\right],$$

making the B-spline combination satisfy $s''(0) = 0$. In the case of a Not-a-Not spline, we similarly need to use that $B'''(x)$ for $x = \{-2, -1, 0, 1, 2\}$ evaluates to $\{1, -2, 0, 2, -1\}$.

B.2.5 Cardinal Splines

Odd-order splines are generally preferred over even-order ones. One reason is illustrated in Figure B.4, showing spline interpolants to *cardinal data*. In contrast to B-splines, cardinal splines (nonzero at just one node) of degree $p > 1$ oscillate forever. However, in the cubic case, the amplitudes decay very rapidly.[7] For successive extrema, the ratio of the magnitudes approach $2 - \sqrt{3} \approx 0.2679$. Increasing the spline smoothness leads to slower decay; for $p = 5, 7, 9$ the ratios become approximately 0.43, 0.53, 0.61, respectively (approaching 1 as $p \to \infty$). Even then, this is benign compared to polynomial cardinal functions (Lagrange polynomials – using no stabilizing end conditions), as illustrated in Figure 1.3. For Gibbs oscillations near to a jump in the data, increasing the spline order from $p = 3$ to $p = \infty$ changes the interpolant from what is shown in Part (c) to Part (b) in Figure 2.3. In the quadratic spline case, the failure of decay can be remedied by letting successive quadratics join half-way between where the data values are imposed.

[7] As to be expected from their minimizing $\int s''(x)^2 dx$ over all possible interpolants.

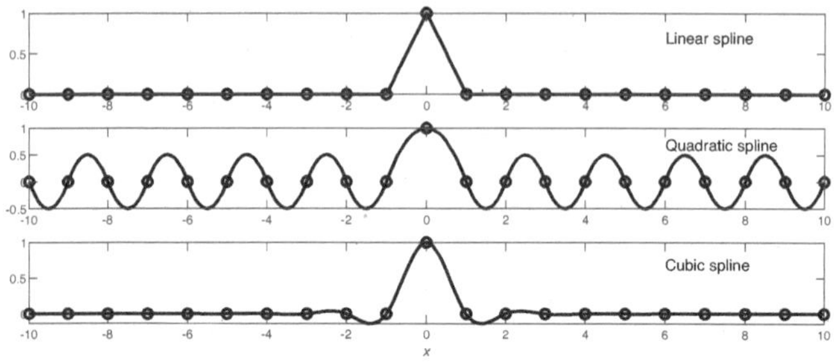

Figure B.4 Equispaced cardinal splines on an infinite interval. Figure reproduced from Fornberg (2021c).

B.2.6 Splines on Nonuniformly Spaced Grids

If the nodes are not equispaced, the procedure in Section B.2.4 still works, as long as we can create nonequispaced cubic B-splines (which will still extend over five nodes). There are three main options for finding these:

(1) piece-by-piece construction of the cubics;
(2) represent $B(x)$ as a linear combination of the functions $|x - x_k|^3$, $k = 1, 2, \ldots, 5$; and
(3) use recursions similar to those in Newton's interpolation formula (detailed, for example, in de Boor (2001, Chapter X) and Powell (1981, Section 19.4)).

For Case (1), one notes that, at a node $x = x_i$, the two cubics that meet can only differ by a term $\alpha_i (x - x_i)^3$. After five nodes, the requirement of starting and ending with all four cubic polynomial coefficients zero leads to a 4×5 homogeneous linear system, guaranteed to have a nontrivial solution.

For Case (2), with equispaced nodes at $x = \{-2, -1, 0, 1, 2\}$,

$$B(x) = \frac{1}{12}\left(1\ |x+2|^3 - 4\ |x+1|^3 + 6\ |x|^3 - 4\ |x-1|^3 + 1\ |x-2|^3\right). \quad \text{(B.3)}$$

This function satisfies all the requirements of a B-spline: It is a cubic in each subinterval, jumps only the third derivative at the nodes, and becomes identically zero outside $[-2, 2]$. This formula generalizes to the case when the five nodes x_i are arbitrarily spaced:

$$B(x) = 2 \sum_{i=1}^{5} \left(\frac{1}{\prod_{k=1, k \neq i}^{5} (x_i - x_k)} \right) |x - x_i|^3, \quad \text{(B.4)}$$

again normalized to make the integral equal to 1.

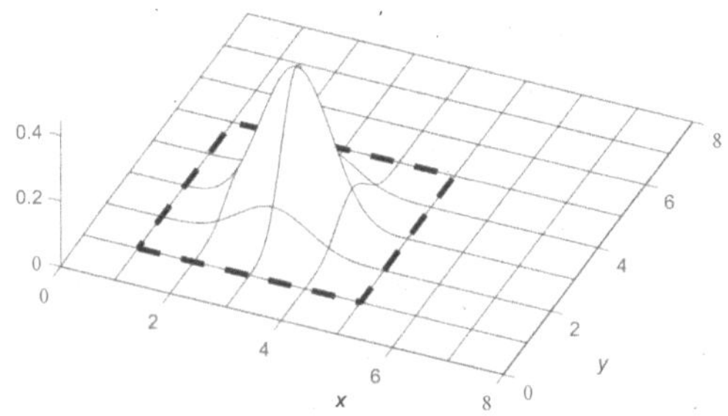

Figure B.5 Tensor product of cubic B-splines in 2-D. Figure reproduced from Fornberg (2021c).

B.3 Generalizations to Multiple Dimensions

B.3.1 Structured Nodes in Multiple Dimensions

Bicubic splines on 2-D rectilinear grids were introduced in de Boor (1962). Higher-dimensional counterparts to the 1-D B-splines are then obtained by forming tensor products, as illustrated in the 2-D case in Figure B.5. Linear combinations of these can then be used similarly to the 1-D illustration in Figure B.3.

B.3.2 Unstructured Nodes in Multiple Dimensions: Radial Basis Functions

The following two observations lead naturally to radial basis functions, which is the topic of Chapter 5. Starting from the present cubic spline perspective, we can note two major opportunities for generalizations:

(1) the interpolation accuracy of splines of order p on a grid with spacing h scales as $O(h^{p+1})$; and
(2) a 1-D function $s(x) = \sum_{k=1}^{N} \lambda_k |x - x_k|^p$ is a spline of order p (assuming p odd[8]; see (B.3), (B.4)). This can be written as

$$s(x) = \sum_{k=1}^{N} \lambda_k \phi(|x - x_k|), \text{ with } \phi(r) = r^{\,p}. \tag{B.5}$$

Regarding Item (1), the order of the error $O(h^{p+1})$ is a consequence of $|r|^p$ (with p odd) having its pth derivative discontinuous at the origin. Choosing instead, say, $\phi(r) = e^{-(\varepsilon r)^2}$ (where $\varepsilon > 0$ is a *shape parameter*), there is no

[8] If p is even, this formula instead gives a single polynomial of degree p.

derivative discontinuity, and the accuracy generally improves from algebraic to spectral (beyond any algebraic order).

Regarding item (2), one can generalize (B.5) to

$$s(\underline{x}) = \sum_{k=1}^{N} \lambda_k \phi(||\underline{x} - \underline{x}_k||), \tag{B.6}$$

where $||\cdot||$ denotes the standard Euclidean 2-norm. Here, the nodes $\underline{x}_k$ can be arbitrarily scattered in a d-dimensional space.

A wealth of theory and practical computational results and enhancements become now available, as summarized in Chapter 5.

Appendix C

Fourier Transforms, Fourier Series, and the FFT Algorithm

Joseph Fourier[1] developed his groundbreaking expansion methods at first as a tool for analyzing heat flow. In spite of initial resistance from his contemporaries for lack of rigor, these expansions are nowadays among the foremost tools in applied mathematics. Table C.1 summarizes the three versions we will describe in more detail. Each case is a transform pair – allowing one to move both ways between physical variables ($u(x)$ and u_j) and transform variables ($\hat{u}(\omega)$ and $\hat{u}_k$).[2] The typical purpose of applying transforms is that certain operations are simpler in one of the spaces than in the other. Each of the three versions in the table is commented on next, followed by a discussion of the Fast Fourier Transform (FFT) algorithm, which is a computationally rapid way to carry out the DFT.

C.1 Fourier Transform (FT)

With use of Euler's formula $e^{ix} = \cos x + i\ \sin x$, the FT can alternatively be formulated without complex numbers (as in Fourier's original treatment):

$$u(x) = \int_0^\infty (a(\omega)\cos\omega x + b(\omega)\sin\omega x)\,d\omega; \quad \begin{cases} a(\omega) = \dfrac{1}{\pi}\displaystyle\int_{-\infty}^{\infty} u(x)\cos\omega x\,dx, \\ b(\omega) = \dfrac{1}{\pi}\displaystyle\int_{-\infty}^{\infty} u(x)\sin\omega x\,dx. \end{cases}$$

Textbooks on Fourier methods and also many on complex variables treat extensively different applications of the FT, including (but not limited to) differential

[1] 1768–1830, French mathematician, physicist, Egyptologist, and politician.

[2] Details in the formulas for the transforms are not consistent in the literature, for example, the choice of sign in the exponents, and the handling of the $\frac{1}{2\pi}$-factor (such as using $\frac{1}{\sqrt{2\pi}}$ for each integral, or placing a factor of 2π in each exponent).

Table C.1 *The three types of Fourier transform/series, together with their inverses, that will be focused on.*

Fourier transform (FT)	
$u(x) = \int_{-\infty}^{\infty} \hat{u}(\omega) e^{i\omega x} d\omega$, $-\infty < x < \infty$	$\hat{u}(\omega) = \frac{1}{2\pi} \int_{-\infty}^{\infty} u(x) e^{-i\omega x} dx$, $-\infty < \omega < \infty$
Fourier series (FS)	
$u(x) = \sum_{k=-\infty}^{\infty} \hat{u}_k e^{ikx}$, $-\pi < x \leq \pi$	$\hat{u}_k = \frac{1}{2\pi} \int_{-\pi}^{\pi} u(x) e^{-ikx} dx$, $k = 0, \pm 1, \pm 2, \ldots$
Discrete Fourier transform (DFT)	
$u_j = \sum_{k=0}^{N-1} \hat{u}_k e^{2\pi ikj/N}$, $j = 0, 1, \ldots, N-1$	$\hat{u}_k = \frac{1}{N} \sum_{j=0}^{N-1} u_j e^{-2\pi ikj/N}$, $k = 0, 1, \ldots, N-1$

equations, and uses of the convolution theorem. We list here just a few key relations:

Theorem C.1.1 *If $f(x)$ has the FT $\hat{f}(\omega)$, then the FT of $\frac{d^k}{dx^k} f(x)$ is $(i\omega)^k \hat{f}(\omega)$ and the FT of $x^k f(x)$ is $(i)^k \frac{d^k}{d\omega^k} \hat{f}(\omega)$.*

Theorem C.1.2 (Parseval's relation) *If $f(x)$ has the FT $\hat{f}(\omega)$, then*

$$\frac{1}{2\pi} \int_{-\infty}^{\infty} |f(x)|^2 dx = \int_{-\infty}^{\infty} |\hat{f}(\omega)|^2 d\omega.$$

Theorem C.1.3 (Convolution theorem) *If $f(x)$ and $g(x)$ have the FT $\hat{f}(\omega)$ and $\hat{g}(\omega)$, resp., then the FT of $\frac{1}{2\pi} \int_{-\infty}^{\infty} f(\xi) g(x - \xi) d\xi$ is $\hat{f}(\omega) \cdot \hat{g}(\omega)$.*

C.2 Fourier Series (FS)

The real-valued counterpart to the complex FS in Table C.1 becomes

$$u(x) = a_0 + \sum_{k=1}^{\infty} a_k \cos kx + \sum_{k=1}^{\infty} b_k \sin kx, \tag{C.1}$$

where

$$\begin{aligned} a_0 &= \tfrac{1}{2\pi} \int_{-\pi}^{\pi} u(x)\, dx, \\ \left.\begin{aligned} a_k &= \tfrac{1}{\pi} \int_{-\pi}^{\pi} u(x) \cos kx\, dx \\ b_k &= \tfrac{1}{\pi} \int_{-\pi}^{\pi} u(x) \sin kx\, dx \end{aligned}\right\} & \quad k = 1, 2, \ldots . \end{aligned} \tag{C.2}$$

One way to arrive at (C.2) is to minimize[3]

$$\int_{-\pi}^{\pi}\left(u(x)-\left(a_0+\sum_{k=1}^{\infty}a_k\cos kx+\sum_{k=1}^{\infty}b_k\sin kx\right)\right)^2 dx.$$

Alternatively, assuming that an expansion of the form (C.1) is possible, multiplying it in turn by $\cos kx$ and $\sin kx$ and integrating over $[-\pi,\pi]$ gives (C.2). Changing the period in the FS from $[-\pi,\pi]$ to $[-L,L]$ and then letting $L\longrightarrow\infty$ provides one way to verify the FT formulas mentioned earlier. As for the FT, we refer to specialized textbooks for discussions of applications.

C.3 Discrete Fourier Transform (DFT)

The transform pair in Table C.1,

$$u_j=\sum_{k=0}^{N-1}\hat{u}_k e^{2\pi ikj/N},\quad j=0,1,\ldots,N-1, \tag{C.3}$$

and

$$\hat{u}_k=\frac{1}{N}\sum_{j=0}^{N-1}u_j e^{-2\pi ikj/N},\quad k=0,1,\ldots,N-1, \tag{C.4}$$

is conveniently written in matrix × vector form:

$$\begin{bmatrix} 1 & 1 & 1 & 1 & \cdots & 1\\ 1 & \omega & \omega^2 & \omega^3 & \cdots & \omega^{N-1}\\ 1 & \omega^2 & \omega^4 & \omega^6 & \cdots & \omega^{2N-2}\\ \vdots & \vdots & & & & \vdots\\ \vdots & \vdots & & & & \vdots\\ 1 & \omega^{N-1} & \cdots & \cdots & \cdots & \omega^{(N-1)^2}\end{bmatrix}\begin{bmatrix}\hat{u}_0\\ \hat{u}_1\\ \hat{u}_2\\ \vdots\\ \vdots\\ \hat{u}_{N-1}\end{bmatrix}=\begin{bmatrix}u_0\\ u_1\\ u_2\\ \vdots\\ \vdots\\ u_{N-1}\end{bmatrix}, \tag{C.5}$$

where ω is the Nth *root of unity*, that is, $\omega=e^{2\pi i/N}$, and similarly:

$$\frac{1}{N}\begin{bmatrix} 1 & 1 & 1 & 1 & \cdots & 1\\ 1 & \omega^{-1} & \omega^{-2} & \omega^{-3} & \cdots & \omega^{-N+1}\\ 1 & \omega^{-2} & \omega^{-4} & \omega^{-6} & \cdots & \omega^{-2N+2}\\ \vdots & \vdots & & & & \vdots\\ \vdots & \vdots & & & & \vdots\\ 1 & \omega^{-N+1} & \cdots & \cdots & \cdots & \omega^{-(N-1)^2}\end{bmatrix}\begin{bmatrix}u_0\\ u_1\\ u_2\\ \vdots\\ \vdots\\ u_{N-1}\end{bmatrix}=\begin{bmatrix}\hat{u}_0\\ \hat{u}_1\\ \hat{u}_2\\ \vdots\\ \vdots\\ \hat{u}_{N-1}\end{bmatrix}. \tag{C.6}$$

[3] Set in turn $\frac{\partial}{\partial a_k}$ and $\frac{\partial}{\partial b_k}$ to zero.

Multiplying out the product of the two matrices (in (C.3) and (C.4), using the formula for a finite geometric progression, noting that $\omega^N = 1$, and including the $\frac{1}{N}$ factor) gives the identity matrix, verifying that the two relations above indeed form a transform pair.

The FFT algorithm (Section C.4) is a very fast way to evaluate the two matrix × vector products.

C.3.1 Discrete Convolution Theorem

One of the most important applications of the FFT algorithm is that it allows *periodic discrete convolutions* to be calculated rapidly. If the three vectors

$$[x_0, x_1, \ldots, x_{N-1}], \qquad [y_0, y_1, \ldots, y_{N-1}], \qquad [z_0, z_1, \ldots, z_{N-1}] \tag{C.7}$$

satisfy

$$\begin{bmatrix} z_0 & z_{N-1} & z_{N-2} & \ddots & z_1 \\ z_1 & z_0 & z_{N-1} & \ddots & z_2 \\ \ddots & \ddots & \ddots & \ddots & \ddots \\ \ddots & \ddots & \ddots & \ddots & \ddots \\ z_{N-1} & z_{N-2} & \ddots & z_1 & z_0 \end{bmatrix} \begin{bmatrix} x_0 \\ x_1 \\ \vdots \\ \vdots \\ x_{N-1} \end{bmatrix} = \begin{bmatrix} y_0 \\ y_1 \\ \vdots \\ \vdots \\ y_{N-1} \end{bmatrix}, \tag{C.8}$$

then their DFT coefficients satisfy

$$\begin{bmatrix} \hat{z}_0 & & & & \\ & \hat{z}_1 & & & \\ & & \ddots & & \\ & & & \ddots & \\ & & & & \hat{z}_{N-1} \end{bmatrix} \begin{bmatrix} \hat{x}_0 \\ \hat{x}_1 \\ \vdots \\ \vdots \\ \hat{x}_{N-1} \end{bmatrix} = \frac{1}{N} \begin{bmatrix} \hat{y}_0 \\ \hat{y}_1 \\ \vdots \\ \vdots \\ \hat{y}_{N-1} \end{bmatrix}. \tag{C.9}$$

Since this equation amounts to a simple multiplication of the DFT coefficients,

$$\hat{z}_n \hat{x}_n = \frac{1}{N} \hat{y}_n,$$

this result, together with the FFT algorithm, offers a very fast way to calculate any one of the vectors in (C.7) if the other two are provided. This theorem can be effective also in nonperiodic settings.

Example C.3.1 Multiply two polynomials of degree N in $O(N \log N)$ operations.

Solution The product of

$$p_1(x) = a_0 + a_1 x + a_2 x^2 + \cdots + a_N x^N$$

and

$$p_2(x) = b_0 + b_1 x + b_2 x^2 + \cdots + b_N x^N$$

is

$$\begin{aligned}
p_1(x) \cdot p_2(x) = \quad & 1 && \cdot \{a_0 b_0\} + \\
& x && \cdot \{a_1 b_0 + a_0 b_1\} + \\
& x^2 && \cdot \{a_2 b_0 + a_1 b_1 + a_0 b_2\} + \\
& \vdots && \\
& x^N && \cdot \{a_N b_0 + a_{N-1} b_1 + a_{N-2} b_2 + \cdots \\
& && \quad + a_1 b_{N-1} + a_0 b_N\} + \\
& \vdots && \\
& x^{2N-1} && \cdot \{a_N b_{N-1} + a_{N-1} b_N\} + \\
& x^{2N} && \cdot \{a_N b_N\} \\
& = && c_0 + c_1 x + c_2 x^2 + \cdots + c_{2N} x^{2N}.
\end{aligned}$$

These coefficient relations can be written as

$$\begin{bmatrix}
a_0 & 0 & \cdots & \cdots & 0 & a_N & \cdots & a_2 & a_1 \\
a_1 & a_0 & 0 & \cdots & \cdots & 0 & a_N & \cdots & a_2 \\
a_2 & a_1 & a_0 & 0 & \cdots & \cdots & 0 & a_N & \cdots \\
\cdots & a_2 & a_1 & a_0 & 0 & \cdots & \cdots & 0 & a_N \\
a_N & \cdots & a_2 & a_1 & a_0 & 0 & \cdots & \cdots & 0 \\
0 & a_N & \cdots & a_2 & a_1 & a_0 & 0 & \cdots & \cdots \\
\cdots & 0 & a_N & \cdots & a_2 & a_1 & a_0 & 0 & \cdots \\
\cdots & \cdots & 0 & a_N & \cdots & a_2 & a_1 & a_0 & 0 \\
0 & \cdots & \cdots & 0 & a_N & \cdots & a_2 & a_1 & a_0
\end{bmatrix}
\begin{bmatrix} b_0 \\ b_1 \\ b_2 \\ \vdots \\ b_N \\ 0 \\ \vdots \\ \vdots \\ 0 \end{bmatrix}
=
\begin{bmatrix} c_0 \\ c_1 \\ c_2 \\ \vdots \\ c_N \\ \vdots \\ \vdots \\ \vdots \\ c_{2N} \end{bmatrix},$$

where both the a- and b-vectors have been padded with extra zeros. This matrix × vector product (to give the c-vector) is exactly of the form that, using the discrete convolution theorem and the FFT algorithm (Section C.4) evaluates in $O(N \log N)$ (instead of $O(N^2)$) operations. □

C.3.2 Aliasing

A second major application of the DFT is to provide trigonometric interpolants to periodic discrete data. As a background to this, one needs to consider *aliasing*.

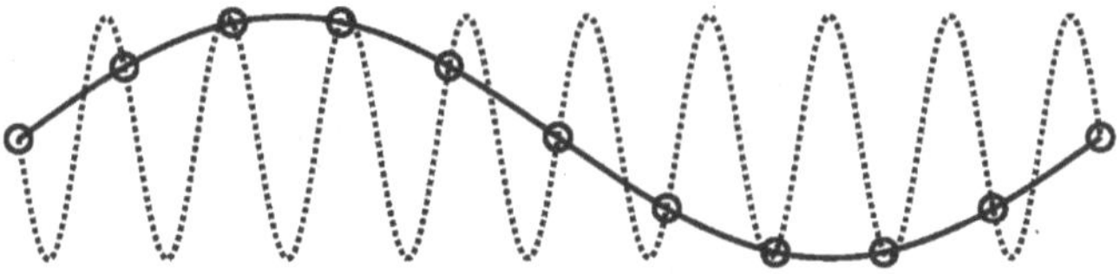

Figure C.1 Example of aliasing: The functions $\sin(1x)$ and $\sin(-9x)$ are indistinguishable at the $N = 10$ grid points $x_j = \frac{2\pi j}{N}$, $j = 0, 1, \ldots, N$ equispaced over $[0, 2\pi]$.

It follows directly from (C.4) that

$$\hat{u}_{k+pN} = \frac{1}{N}\sum_{j=0}^{N-1} u_j e^{-2\pi i(k+pN)j/N} = \hat{u}_k$$

for any integer p. Thus, modes for which k-values differ by N (or by any multiple of N) become indistinguishable at the node points, as illustrated in Figure C.1.

C.3.3 Relation between Fourier Series and Discrete Fourier Transform Coefficients

We next consider a $[-\pi, \pi]$-periodic function $u(x)$, with the Fourier series

$$u(x) = \sum_{\nu=-\infty}^{\infty} \hat{u}_\nu e^{i\nu x}. \tag{C.10}$$

Sampling this function at the discrete points $x_j = j\frac{2\pi}{N}$, $j = 0, 1, \ldots, N-1$ (which are equispaced within the $[0, 2\pi]$-period), we denote the resulting DFT coefficients by $\hat{U}_k$ (to distinguish from the Fourier series coefficients $\hat{u}_\nu$). These two coefficient sets are closely related. Substituting (C.10) into the expression for $\hat{U}_k$ gives

$$\begin{aligned}\hat{U}_k &= \frac{1}{N}\sum_{j=0}^{N-1} u_j e^{-2\pi ikj/N} = \frac{1}{N}\sum_{j=0}^{N-1}\left(\sum_{\nu=-\infty}^{\infty}\hat{u}_\nu e^{2\pi ij\nu/N}\right)e^{-2\pi ijk/N}\\ &= \sum_{\nu=-\infty}^{\infty}\hat{u}_\nu\left(\frac{1}{N}\sum_{j=0}^{N-1}e^{2\pi ij(\nu-k)/N}\right) = \sum_{n=-\infty}^{\infty}\hat{u}_{k+nN}.\end{aligned} \tag{C.11}$$

The effect of sampling $u(x)$ at the equispaced discrete points therefore leads to an *aliasing* related effect – all the modes that cannot be resolved by the grid are "folded back" (aliased) onto those that are resolved by the grid.

This relation, $\hat{U}_k = \sum_{n=-\infty}^{\infty} \hat{u}_{k+nN}$, offers a practical tool to computationally approximate the leading Fourier coefficients $\hat{u}_k$ of a 2π-periodic function. One simply chooses N sufficiently large that $\hat{u}_{k+nN}$ is dominated by the $n = 0$ case, and then compute the DFT coefficients $\hat{U}_k$ with the FFT algorithm.

C.3.4 Numbering of the Modes in Connection with Trigonometric Interpolation

This issue is a common cause of programming errors when utilizing the FFT for tasks that are mathematically described by a Fourier series or a Fourier transform. FFTs work with Fourier modes e^{ikx}, for $k = 0, 1, 2, \ldots, N-1$ (one-sided in k) whereas mathematical descriptions mostly use k-ranges centered around $k = 0$. Let us for now assume that N is odd (slightly simplifying the algebra; Section C.3.6 describes equispaced interpolation for N either even or odd). Suppose we are given $u_0, u_1, \ldots, u_{N-1}$, all real numbers, and that we have calculated the complex DFT coefficients $\hat{u}_0, \hat{u}_1, \ldots, \hat{u}_{N-1}$ using (C.6). At the node locations $x_j = 2\pi j/N$, we can then evaluate

$$\sum_{k=0}^{N-1} \hat{u}_k e^{ikx_j},$$

which will return the values u_j, for $j = 0, 1, \ldots, N-1$ that we started with. It then seems reasonable that, if we evaluate this sum for values of x in between the x_j:

$$U(x) = \sum_{k=0}^{N-1} \hat{u}_k e^{ikx}, \tag{C.12}$$

we will obtain interpolated values. However, this leads to a nasty surprise. One manifestation of this is that, even when all the original data is real, this interpolant becomes complex valued between the node points ($j = 0, 1, \ldots, N-1$).

Figure C.2(a) shows as circles an example of purely real data (with crosses showing their imaginary parts to be zero), and in solid and dashed curves the real and imaginary parts, respectively, of the Fourier interpolant when using (C.12). The mode renumbering that will be described next gives instead the interpolant shown in part (b).

Mode renumbering: To resolve this, we need to make use of *aliasing*:

Fourier modes	$\hat{u}_0$	$\hat{u}_1$	$\cdots$	$\hat{u}_{\frac{N-1}{2}}$	$\hat{u}_{\frac{N+1}{2}}$	$\cdots$	$\hat{u}_{N-2}$	$\hat{u}_{N-1}$
Corresponding to wave numbers	0	1	$\cdots$	$\frac{N-1}{2}$	$\frac{N+1}{2}$	$\cdots$	$N-2$	$N-1$
By aliasing, can instead be viewed as wave numbers	0	1	$\cdots$	$\frac{N-1}{2}$	$-\frac{N-1}{2}$	$\cdots$	-2	-1

For the right-hand half of the modes, we need to reinterpret them from modes $\frac{N+1}{2}, \ldots, N-2, N-1$ to modes $-\frac{N-1}{2}, \ldots, -2, -1$. The discrete transform (C.3) can thus alternatively be written as

$$u_j = \sum_{k=-\frac{1}{2}(N-1)}^{\frac{1}{2}(N-1)} \hat{u}_k e^{2\pi i jk/N},$$

with the Fourier interpolant,

$$U(x) = \sum_{k=-\frac{1}{2}(N-1)}^{\frac{1}{2}(N-1)} \hat{u}_k e^{ikx}. \tag{C.13}$$

Figures C.2(a), (b) were obtained using (C.12) and (C.13), respectively.

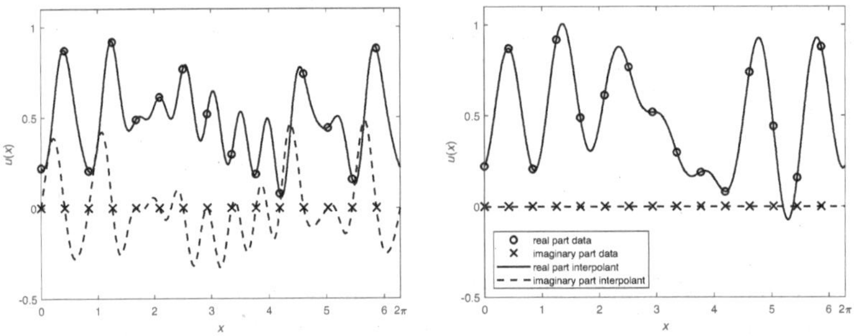

(a) Fourier interpolant (C.12)

(b) Same data, but with the modes handled correctly

Figure C.2 Fourier interpolation of real data (circles and crosses); (a) without; and (b) with correct interpretation of the Fourier modes.

If the original function u_j is real, then $\hat{u}_0$ is real and $\hat{u}_{-k} = \overline{\hat{u}_k}$, with the result that $U(x) = \overline{U(x)}$, that is, $U(x)$ is a real function (as it should be).

If N is even, an additional slight complication arises. From the aliasing result

(C.11) follows that

$$\hat{u}_{-\frac{1}{2}N} = \hat{u}_{\frac{1}{2}N}.$$

This is known as the *reflection* or *Nyquist frequency* and requires the following modification in the interpolation formula,

$$U(x) = \sum_{k=-\frac{1}{2}N+1}^{\frac{1}{2}N-1} \hat{u}_k e^{ikx} + \hat{u}_{\pm\frac{1}{2}N} \cos\left(\frac{1}{2}Nx\right),$$

where $\hat{u}_{\pm\frac{1}{2}N} := \hat{u}_{+\frac{1}{2}N} = \hat{u}_{-\frac{1}{2}N}$. In many cases, one can simply set $\hat{u}_{\pm\frac{1}{2}N}$ to zero (or just ignore its presence, depending on the application).

C.3.5 Character of Transform for Different Cases of Input Data

If the N input numbers to a DFT are complex, so are in general the output numbers. This is illustrated as the top case in Figure C.3. If the input is a real vector (next case in Figure C.3), the output will have the particular structure illustrated to the right. Specialized codes are available for this case, using as input and delivering as output only the entries inside the boxes shown with thick boundary lines. The last two cases – cosine and sine transforms – also arise often in applications. The real input data is here symmetric or antisymmetric, and the transforms take on a similar structure. These examples illustrate again that it is necessary to interpret the modes in the order $[0, 1, 2, 3, \ldots, -3, -2, -1]$, as just described (rather than $[0, 1, 2, 3, \ldots, N-3, N-2, N-1]$).

In this Figure C.3, N has been assumed to be even. This is traditionally a common case in computations because the FFT algorithm is particularly simple for sizes (values of N) that are powers of two. One then has to keep in mind that the mode $N/2$ then does not quite "fit the pattern," as pointed out earlier.

C.3.6 FFT-based Interpolation

The FFT algorithm (described next in Section C.4) provides a computationally rapid means for operating on equispaced periodic discrete data for tasks such as differentiation and interpolation. Codes using the FFT need to take into account the data arrangement shown in Figure C.3. As an example, the code shown in Algorithm C.1 performs FFT-based, equispaced periodic interpolation.

Figure C.4 provides an example of its usage, in the case of $N = 16$ input values in the f-vector and $M = 50$ output values, to be placed in the g-vector. The circles show the input values (here obtained by `rng(0); f=randn(1,N);`) placed at x-coordinates $0, 1, \ldots, N-1$ (with the value at N assumed to repeat

Transform type	Structure of Input	Structure of Output
Complex	General, complex Re: $a_0\ \ a_1\ \cdots\ \cdots\ a_{\frac{N}{2}-1}\ \ a_{\frac{N}{2}}\ \ a_{\frac{N}{2}+1}\ \cdots\ \cdots\ a_{N-1}$ Im: $b_0\ \ b_1\ \cdots\ \cdots\ b_{\frac{N}{2}-1}\ \ b_{\frac{N}{2}}\ \ b_{\frac{N}{2}+1}\ \cdots\ \cdots\ b_{N-1}$	General, complex Re: $c_0\ \ c_1\ \cdots\ \cdots\ c_{\frac{N}{2}-1}\ \ c_{\frac{N}{2}}\ \ c_{\frac{N}{2}+1}\ \cdots\ \cdots\ c_{N-1}$ Im: $d_0\ \ d_1\ \cdots\ \cdots\ d_{\frac{N}{2}-1}\ \ d_{\frac{N}{2}}\ \ d_{\frac{N}{2}+1}\ \cdots\ \cdots\ d_{N-1}$
Real	Real Re: $a_0\ \ a_1\ \cdots\ \cdots\ a_{\frac{N}{2}-1}\ \ a_{\frac{N}{2}}\ \ a_{\frac{N}{2}+1}\ \cdots\ \cdots\ a_{N-1}$ Im: $0\ \ 0\ \cdots\ \cdots\ 0\ \ 0\ \ 0\ \cdots\ \cdots\ 0$	Real part symmetric, imaginary part antisymmetric Re: $c_0\ \ c_1\ \cdots\ \cdots\ c_{\frac{N}{2}-1}\ \ c_{\frac{N}{2}}\ \ c_{\frac{N}{2}-1}\ \cdots\ \cdots\ c_1$ Im: $0\ \ d_1\ \cdots\ \cdots\ d_{\frac{N}{2}-1}\ \ 0\ \ -d_{\frac{N}{2}-1}\ \cdots\ \cdots\ -d_1$
Cosine	Real , symmetric Re: $a_0\ \ a_1\ \cdots\ \cdots\ a_{\frac{N}{2}-1}\ \ a_{\frac{N}{2}}\ \ a_{\frac{N}{2}-1}\ \cdots\ \cdots\ a_1$ Im: $0\ \ 0\ \cdots\ \cdots\ 0\ \ 0\ \ 0\ \cdots\ \cdots\ 0$	Real, symmetric Re: $c_0\ \ c_1\ \cdots\ \cdots\ c_{\frac{N}{2}-1}\ \ c_{\frac{N}{2}}\ \ c_{\frac{N}{2}-1}\ \cdots\ \cdots\ c_1$ Im: $0\ \ 0\ \cdots\ \cdots\ 0\ \ 0\ \ 0\ \cdots\ \cdots\ 0$
Sine	Real, antisymmetric Re: $0\ \ a_1\ \cdots\ \cdots\ a_{\frac{N}{2}-1}\ \ 0\ \ -a_{\frac{N}{2}-1}\ \cdots\ \cdots\ -a_1$ Im: $0\ \ 0\ \cdots\ \cdots\ 0\ \ 0\ \ 0\ \cdots\ \cdots\ 0$	Imaginary, antisymmetric Re: $0\ \ 0\ \cdots\ \cdots\ 0\ \ 0\ \ 0\ \cdots\ \cdots\ 0$ Im: $0\ \ d_1\ \cdots\ \cdots\ d_{\frac{N}{2}-1}\ \ 0\ \ -d_{\frac{N}{2}-1}\ \cdots\ \cdots\ -d_1$

Figure C.3 Comparison between a general complex DFT and some special cases of it in the case of N even. The heavy lines surround the data vectors which specialized routines for the cases typically use as input and output (other table entries are superfluous).

Algorithm C.1 Code for equispaced periodic FFT-based interpolation.

```
function g = fourier_interp(f,M)
% Input parameters
%     f   Vector (row or column) with function values at x = 0,1,2,...,N-1
%         (treated as N-periodic)
%     M   Number of interpolated values to be returned; N and M
%         can be either even or odd.
% Output parameter
%     g   Column vector with the interpolated values at locations
%         [0,1,2,...,M-1]*N/M.

N = length(f); n = floor(N/2);
fh = ifft(f);           % Transform from physical to Fourier space
gh = zeros(M,1);        % Vector for storing updated Fourier coefficients
gh(1) = fh(1);
f = ones(n,1); if mod(N,2)==0; f(n) = 0.5; end
for k = 1:n             % Prepare vector for return to physical space
    i1 = 1 + mod( k,M);
    i2 = M - mod(k-1,M);
    gh(i1) = gh(i1) + f(k)*fh(k+1);
    gh(i2) = gh(i2) + f(k)*fh(N+1-k);
end
g = fft(gh);           % Transform back to physical space
end
```

the value at 0). The code produces in this case $M = 50$ equispaced interpolated values, shown as medium-sized dots at locations $(0, 1, \ldots, M-1)*N/M$ (again, with the value at N assumed to match the one at 0). The small dots along the horizontal axis confirm that the imaginary part becomes zero (in fact, to machine precision) at all these M locations.[4] The continuous curve is obtained by using a large M-value. Whatever value of M is used, the interpolated points will fall on this curve (representing the lowest degree interpolating trigonometric polynomial to the original data).

C.4 Fast Fourier Transform (FFT)

C.4.1 Historical Background

The discovery by James W. Cooley and John W. Tukey in 1965 of the FFT algorithm caused a computational revolution. Applications of the FFT soon proved abundant in nearly all fields. Not only could many existing tasks be solved orders of magnitude faster but computing could also be brought to bear in new areas. The FFT principle has since been found in several earlier works,

[4] The computational efficiency for real-valued input data can be doubled if one places a second interpolation case in the imaginary part of the input vector. Since the interpolation process is linear in its input data, the result for this second case is obtained in the imaginary part of the output vector.

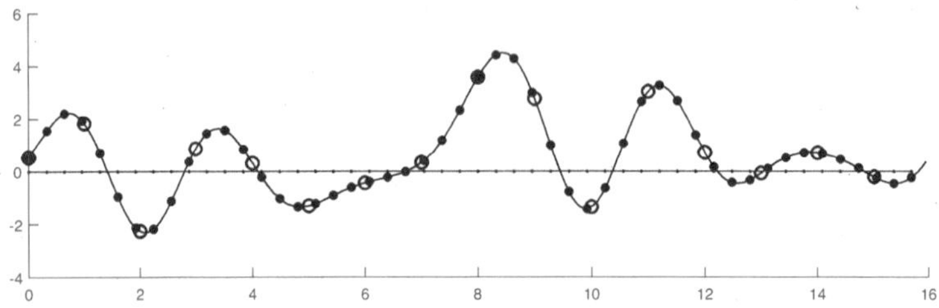

Figure C.4 FFT-based interpolation, as described in Section C.3.6.

$$\begin{bmatrix}
\omega^0 & \omega^0 & \omega^0 & \omega^0 & \omega^0 & \omega^0 & \omega^0 & \omega^0 \\
\omega^0 & \omega^1 & \omega^2 & \omega^3 & \omega^4 & \omega^5 & \omega^6 & \omega^7 \\
\omega^0 & \omega^2 & \omega^4 & \omega^6 & \omega^8 & \omega^{10} & \omega^{12} & \omega^{14} \\
\omega^0 & \omega^3 & \omega^6 & \omega^9 & \omega^{12} & \omega^{15} & \omega^{18} & \omega^{21} \\
\omega^0 & \omega^4 & \omega^8 & \omega^{12} & \omega^{16} & \omega^{20} & \omega^{24} & \omega^{28} \\
\omega^0 & \omega^5 & \omega^{10} & \omega^{15} & \omega^{20} & \omega^{25} & \omega^{30} & \omega^{35} \\
\omega^0 & \omega^6 & \omega^{12} & \omega^{18} & \omega^{24} & \omega^{30} & \omega^{36} & \omega^{42} \\
\omega^0 & \omega^7 & \omega^{14} & \omega^{21} & \omega^{28} & \omega^{35} & \omega^{42} & \omega^{49}
\end{bmatrix}
=
\left[\begin{array}{cccc|cccc}
\omega^0 & & & & \omega^0 & & & \\
& \omega^0 & & & & \omega^1 & & \\
& & \omega^0 & & & & \omega^2 & \\
& & & \omega^0 & & & & \omega^3 \\
\hline
\omega^0 & & & & -\omega^0 & & & \\
& \omega^0 & & & & -\omega^1 & & \\
& & \omega^0 & & & & -\omega^2 & \\
& & & \omega^0 & & & & -\omega^3
\end{array}\right]
\times$$

$$\times
\left[\begin{array}{cccc|cccc}
\omega^0 & \omega^0 & \omega^0 & \omega^0 & & & & \\
\omega^0 & \omega^2 & \omega^4 & \omega^6 & & & & \\
\omega^0 & \omega^4 & \omega^8 & \omega^{12} & & & & \\
\omega^0 & \omega^6 & \omega^{12} & \omega^{18} & & & & \\
\hline
& & & & \omega^0 & \omega^0 & \omega^0 & \omega^0 \\
& & & & \omega^0 & \omega^2 & \omega^4 & \omega^6 \\
& & & & \omega^0 & \omega^4 & \omega^8 & \omega^{12} \\
& & & & \omega^0 & \omega^6 & \omega^{12} & \omega^{18}
\end{array}\right]
\times
\begin{bmatrix}
1 & & & & & & & \\
& & 1 & & & & & \\
& & & & 1 & & & \\
& & & & & & 1 & \\
& 1 & & & & & & \\
& & & 1 & & & & \\
& & & & & 1 & & \\
& & & & & & & 1
\end{bmatrix}$$

Figure C.5 The first step in the FFT factorization of the DFT matrix, with $\omega = e^{2\pi i/N}$ and shown for $N = 8$.

for example, by Gauss in 1805 and Runge in 1903 and 1905. It is described in a numerical survey book by Runge and König in 1924 and again in a 1939 book on trigonometric computations by Stumpff. X-ray crystallographers in Cambridge used the method in the 1930s. It was also described in this context by Danielson and Lanczos in 1942. Still, it remained on the fringes of numerical knowledge until its revolutionary potential became apparent to the rediscoverers, Cooley and Tukey.

Part of the reason the FFT idea had not had much impact earlier was that electronic computers were then not available. Clever symmetries had been found that were just as effective in speeding up the calculations for the small problem sizes that were all that then could be handled.

C.4.2 FFT Algorithm

It appears from (C.5) and (C.6) that each of these matrix $\times$ vector multiplications will require $O(N^2)$ arithmetic operations. However, the matrices have simple patterns in their elements, and do not depend on the data to be transformed. An easy way to visualize the FFT concept is to note that the matrices can be written as a product of $O(\log N)$ matrices with each only $O(N)$ nonzero entries. Applying these sparse matrices in turn to a data vector reduces the operation count to $O(N \log N)$ – a vast gain over $O(N^2)$ when N is large. As every scientific programming environment has highly optimized FFT codes available, users of FFTs rarely (if ever) write their own codes. We limit ourselves here to illustrate matrix factorization in the case of an 8×8 DFT matrix (which suffices to show the FFT principle in cases of N a power of two).

The original Cooley–Tukey factorization contains as its key step the factorization shown in Figure C.5. The 8×8 DFT matrix is split into three factors: From the left (i) a very sparse matrix with just two entries per row, (ii) a block 2×2 matrix where the two nontrivial (diagonal) blocks are again perfect DFT matrices, and (iii) a permutation matrix. Repeating the same DFT matrix split-up on the middle factor splits similarly out another sparse factor to the left and another permutation matrix to the right, with the remaining central factor having four 2×2 DFT blocks along its diagonal. After a third such step (note that $3 = \log_2 8$), the central part has become the identity matrix, there are three sparse matrix factors to the left, and to the right a permutation matrix (since the product of any number of permutation matrices is again a permutation matrix). This gives rise to the factorization denoted "Cooley–Tukey" in Figure C.6. Although there turns out to be a simple ("bit-reversal") pattern in the entries of the permutation matrix and it can be created very quickly, it can also be convenient to eliminate this matrix altogether in exchange for slightly different main factors. One such version is in the figure denoted "Glassman" (Glassman, 1970).

If the total number of nodes N is not a power of 2 but a product of small factors, similar matrix factorizations are again available. In the "worst case" of N being a prime, the constant in the $O(N \log N)$ operation count need not increase by more than about 20% (Winograd, 1978). Generalizations of the FFT to non-equispaced data can be found with operation counts $O(N \log N + N \log(1/\varepsilon))$ where ε is the desired precision (Dutt and Rokhlin, 1993; Greengard and Lee, 2004).

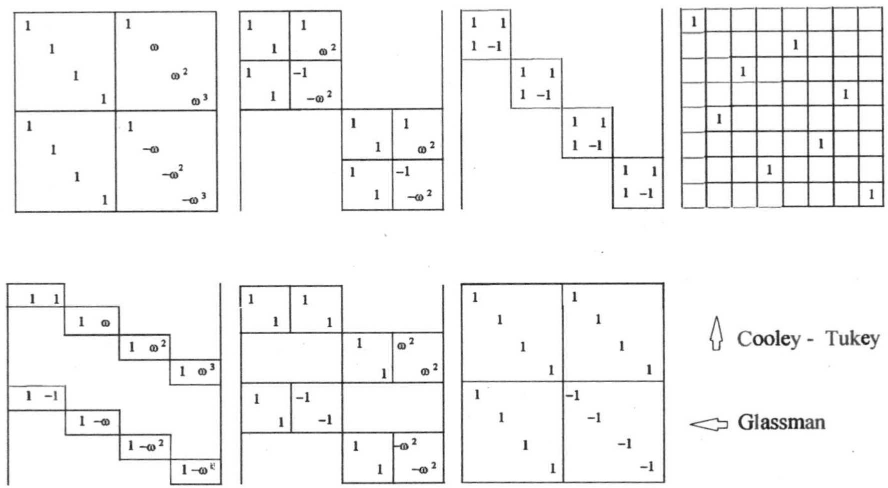

Figure C.6 The structure of two different sparse factorizations of the 8 × 8 DFT matrix shown in Figure C.5.

Appendix D
Lagrange Multipliers

Lagrange multipliers provide a tool for both analytical and numerical constrained optimization. To provide background and notation, we first consider the unconstrained case.

D.1 Unconstrained Case

Task: Find local equilibrium points of a scalar function $f(x_1, x_2, \ldots, x_n)$.[1]

Procedure: Solving the $n \times n$ system $\nabla f = \underline{0}$ gives all equilibrium points. Here, ∇ denotes the gradient, that is, the system $\nabla f = \underline{0}$ can be written out as $\frac{\partial f}{\partial x_1} = 0, \frac{\partial f}{\partial x_2} = 0, \ldots, \frac{\partial f}{\partial x_n} = 0.$

D.2 Constrained Case: The Lagrange Multiplier Procedure

Task: Find the local equilibrium points of the scalar function

$$f(x_1, x_2, \ldots, x_n)$$

subject to the constraints

$$\begin{cases} g_1(x_1, x_2, \ldots, x_n) & = 0 \\ g_2(x_1, x_2, \ldots, x_n) & = 0 \\ \quad\vdots & \;\;\vdots \\ g_k(x_1, x_2, \ldots, x_n) & = 0 \end{cases} \qquad (k < n). \tag{D.1}$$

[1] A separate step may be needed to then determine if an equilibrium point is a local extremal point and, if so, of what type.

Procedure: Form

$$h = f + (\lambda_1 g_1 + \lambda_2 g_2 + \cdots + \lambda_k g_k) \tag{D.2}$$

and

$$\nabla h = \underline{0}. \tag{D.3}$$

The relation (D.3) amounts to n equations in $n + k$ unknowns:

$$x_1, x_2, \ldots, x_n, \lambda_1, \ldots, \lambda_k.$$

Together with the k constraints (D.1), this becomes $n + k$ equations in $n + k$ unknowns. In the solution, the λ-coefficients are normally discarded.

Proof 1. *Characterize infinitesimal permissible node movements*

$$[\Delta x_1, \Delta x_2, \ldots, \Delta x_n]$$

which obey the constraints.

Focusing on the constraint equations: If we are at some location $x_1, x_2, \ldots, x_n$ which satisfies (D.1) and move infinitesimal distances $\Delta x_1, \Delta x_2, \ldots, \Delta x_n$, these distances must, due to (D.1), satisfy

$$\begin{cases} \frac{\partial g_1}{\partial x_1}\Delta x_1 + \frac{\partial g_1}{\partial x_2}\Delta x_2 + \ldots + \frac{\partial g_1}{\partial x_n}\Delta x_n & = 0, \\ \vdots & \vdots \\ \frac{\partial g_k}{\partial x_1}\Delta x_1 + \frac{\partial g_k}{\partial x_2}\Delta x_2 + \ldots + \frac{\partial g_k}{\partial x_n}\Delta x_n & = 0. \end{cases}$$

This can be written as

$$[\Delta x_1, \Delta x_2, \ldots, \Delta x_n] \begin{bmatrix} \frac{\partial g_1}{\partial x_1} & \cdots & \frac{\partial g_k}{\partial x_1} \\ \frac{\partial g_1}{\partial x_2} & \cdots & \frac{\partial g_k}{\partial x_2} \\ \vdots & & \vdots \\ \frac{\partial g_1}{\partial x_n} & \cdots & \frac{\partial g_k}{\partial x_n} \end{bmatrix} = \begin{bmatrix} 0 & 0 & \cdots & 0 \end{bmatrix}. \tag{D.4}$$

2. *Next, write out the Lagrange multiplier formula* (D.3) *more explicitly:*

$$-\begin{bmatrix} \frac{\partial g_1}{\partial x_1} & \cdots & \frac{\partial g_k}{\partial x_1} \\ \frac{\partial g_1}{\partial x_2} & \cdots & \frac{\partial g_k}{\partial x_2} \\ \vdots & & \vdots \\ \frac{\partial g_1}{\partial x_n} & \cdots & \frac{\partial g_k}{\partial x_n} \end{bmatrix} \begin{bmatrix} \lambda_1 \\ \vdots \\ \lambda_k \end{bmatrix} = \begin{bmatrix} \frac{\partial f}{\partial x_1} \\ \frac{\partial f}{\partial x_2} \\ \vdots \\ \frac{\partial f}{\partial x_n} \end{bmatrix}. \tag{D.5}$$

Multiplying (D.5) by $[\Delta x_1, \Delta x_2, \ldots, \Delta x_n]$ from the left gives, in view of (D.4) (with the identical matrix),

$$\frac{\partial f}{\partial x_1}\Delta x_1 + \frac{\partial f}{\partial x_2}\Delta x_2 + \cdots + \frac{\partial f}{\partial x_n}\Delta x_n = 0.$$

Hence, the point(s) obtained by solving (D.3) together with the constraint equations have the property of equilibrium points, that is, moving away from them in a way consistent with the constraint equations (D.1) leaves, to leading order, the value of $f(x_1, x_2, \ldots, x_n)$ unchanged. □

D.3 Application of Lagrange Multipliers to Constrained Linear Systems

Task: Find the least square solution to the linear system

$$\begin{bmatrix} & A & \end{bmatrix}_{m,n} \begin{bmatrix} \underline{x} \end{bmatrix}_{n,1} = \begin{bmatrix} \underline{b} \end{bmatrix}_{m,1}, \tag{D.6}$$

that is, a vector $\underline{x}$ which minimizes $||A\underline{x} - \underline{b}||_2^2 = (A\underline{x} - \underline{b})^T(A\underline{x} - \underline{b})$ subject to linear constraint equations

$$\begin{bmatrix} & C & \end{bmatrix}_{k,n} \begin{bmatrix} \underline{x} \end{bmatrix}_{n,1} = \begin{bmatrix} \underline{c} \end{bmatrix}_{k,1}, \tag{D.7}$$

which need to be enforced exactly. With the indicated matrix sizes, we assume $m + k \geq n$, so there are altogether at least as many equations as unknowns, and $k \leq n$, so there are not more constraints than unknowns.

Procedure: The solution to the constrained least squares problem (D.6), (D.7) is obtained by solving the square linear system

$$\left[\begin{array}{c|c} A^T A & C^T \\ \hline C & 0 \end{array}\right] \left[\begin{array}{c} \underline{x} \\ \hline \underline{\lambda} \end{array}\right] = \left[\begin{array}{c} A^T \underline{b} \\ \hline \underline{c} \end{array}\right]. \tag{D.8}$$

The $\underline{x}$-part in the solution vector provides the solution to the problem. The bottom block of equations in (D.8) ensures that the constraints indeed are satisfied exactly. We also note that, if there are no constraint equations, (D.8) reduces to the standard normal equations for solving a constraint-free overdetermined linear system (D.6) with $m > n$. Numerical aspects on solving large systems of the form (D.6), (D.7) are discussed in Scott and Tůma (2022).

Proof The task is to minimize

$$f(x_1, x_2, \ldots, x_n) = \frac{1}{2}(A\underline{x} - \underline{b})^T(A\underline{x} - \underline{b}) = \left\{\frac{1}{2}\underline{x}^T A^T A\underline{x} - \underline{x}^T A^T \underline{b}\right\} + \left\{\frac{1}{2}\underline{b}^T \underline{b}\right\}.$$

Since the last term in the RHS $\left\{\frac{1}{2}\underline{b}^T\underline{b}\right\}$ is a constant, we will omit it. More explicitly, the RHS can then be written as

$$f(x_1, x_2, \ldots, x_n)$$
$$= \frac{1}{2}\,[x_1\, x_2\, \cdots\, x_n]\,\left[A^T A\right]\begin{bmatrix} x_1 \\ x_2 \\ \vdots \\ x_n \end{bmatrix} - [x_1\, x_2\, \ldots\, x_n]\,\left[A^T\right]\begin{bmatrix} b_1 \\ b_2 \\ \vdots \\ b_n \end{bmatrix}, \tag{D.9}$$

from which follows (for $i = 1, 2, \ldots, n$)

$$\frac{\partial f}{\partial x_i} = \left[\text{row } i \text{ of } A^T A\right]\begin{bmatrix} x_1 \\ x_2 \\ \vdots \\ x_n \end{bmatrix} - \left[\text{row } i \text{ of } A^T\right]\begin{bmatrix} b_1 \\ b_2 \\ \vdots \\ b_n \end{bmatrix}. \tag{D.10}$$

Also

$$\frac{\partial}{\partial x_i}(\lambda_1 g_1 + \lambda_2 g_2 + \cdots + \lambda_k g_k) = \lambda_1 \frac{\partial g_1}{\partial x_i} + \lambda_2 \frac{\partial g_2}{\partial x_i} + \cdots + \lambda_1 \frac{\partial g_n}{\partial x_i}$$
$$= \left[\text{row } i \text{ of } C^T\right]\begin{bmatrix} \lambda_1 \\ \lambda_2 \\ \vdots \\ \lambda_k \end{bmatrix}. \tag{D.11}$$

Thus (with h defined in (D.2))

$$\frac{\partial h}{\partial x_i} = \{\text{sum of the RHSs of (D.10) and (D.11)}\}.$$

For $i = 1, 2, \ldots, n$, this gives the top n rows of (D.8). As noted earlier, the bottom k rows of (D.8) are made up of the constraint equations. □

Corollary *If the matrix A is positive definite (implying square and symmetric), one can form $A^{-1/2}$, also positive definite. Then, with the constraints* (D.7), $||A^{-1/2}(A\underline{x} - \underline{b})||_2^2 = (A\underline{x} - \underline{b})^T A^{-1}(A\underline{x} - \underline{b})$ *is minimized by $\underline{x}$ obtained by solving*

$$\left[\begin{array}{c|c} A & C^T \\ \hline C & 0 \end{array}\right]\left[\begin{array}{c} \underline{x} \\ \hline \underline{\lambda} \end{array}\right] = \left[\begin{array}{c} \underline{b} \\ \hline \underline{c} \end{array}\right]. \tag{D.12}$$

The proof is similar to the one mentioned earlier, and we omit it here. In the case A is negative definite, the $\underline{x}$-vector obtained from (D.12) similarly minimizes $||(-A)^{-1/2}(A\underline{x} - \underline{b})||_2^2$.

Appendix E
Extrapolation Methods

Although extrapolation intrinsically is a less stable task than interpolation, it can nevertheless be highly effective. In the context of FD methods, a common situation is to have numerical results for a sequence of h (step size) values approaching $h = 0$, together with theoretical knowledge that the computed solution (point-wise) possesses a convergent Taylor expansion (with unknown coefficients) around $h = 0$. Extrapolation down to $h = 0$ is then well posed and can be carried out by Richardson extrapolation (Section E.1). Deferred correction can sometimes (especially in connection with ODEs and PDEs) provide several additional orders of accuracy from a lower-order scheme (Section 3.3.2).

The two methods just mentioned can be characterized as linear. In other applications, such as accelerating slowly converging iterations and summations, nonlinear techniques are often more effective, as surveyed in several works (Brezinski, 2009; Brezinski and Zaglia, 1991; Sidi, 2003). One such method, Aitken extrapolation, is briefly described in Section E.2. Conversion of truncated Taylor expansions to Padé rational form can, in certain contexts (especially related to analytic functions), provide spectacularly effective extrapolations, as discussed in Section E.3.

E.1 Richardson Extrapolation

When using FD formulas in simple geometries, as well as when evaluating integrals with standard methods for equispaced grids, the resulting errors typically take the form of a power series in the discretization step size h, in many cases with only even powers of h present.[1] With the approximation obtained with step size h denoted by $\mathrm{A}(h)$ and the exact (unknown) result by E, one has

[1] This situation arises also for the ODE extrapolation method described in Section 3.1.4 and again for the parallel-in-time method described in Section F.4.

in this case

$$A(h) = E + c_1 h^2 + c_2 h^4 + c_3 h^6 + \cdots , \tag{E.1}$$

where the c-coefficients are also unknown. The next two sections describe a couple of common ways to proceed. More information on Richardson extrapolation can be found in Zlatev et al. (2018) (in particular with regard to its application to time stepping systems of ODEs).

E.1.1 Successively Halving the Step Size

Using in (E.1) $h/2$ in place of h gives

$$A\left(\frac{h}{2}\right) = E + c_1 \frac{1}{4} h^2 + c_2 \frac{1}{16} h^4 + c_3 \frac{1}{64} h^6 + \cdots , \tag{E.2}$$

thus (after eliminating c_1 between (E.1) and (E.2))

$$E = \frac{4A(\frac{h}{2}) - A(h)}{3} + d_2 h^4 + d_3 h^6 + \cdots , \tag{E.3}$$

which is fourth-order accurate, compared to the second-order $E = A(h) - c_1 h^2 - c_2 h^4 - \cdots$ obtained directly from (E.1). This concept is readily repeated and then represented in a *Romberg table*[2]

$$\begin{array}{llllllll}
A(h) & \searrow & & & & & & \\
A(\frac{h}{2}) & \overset{\rightarrow}{\searrow} & B(\frac{h}{2}) = \frac{4A(\frac{h}{2})-A(h)}{3} & \searrow & & & & \\
A(\frac{h}{4}) & \overset{\rightarrow}{\searrow} & B(\frac{h}{4}) = \frac{4A(\frac{h}{4})-A(\frac{h}{2})}{3} & \overset{\rightarrow}{\searrow} & C(\frac{h}{4}) = \frac{16B(\frac{h}{4})-B(\frac{h}{2})}{15} & \searrow & & \\
A(\frac{h}{8}) & \overset{\rightarrow}{\searrow} & B(\frac{h}{8}) = \frac{4A(\frac{h}{8})-A(\frac{h}{4})}{3} & \overset{\rightarrow}{\searrow} & C(\frac{h}{8}) = \frac{16B(\frac{h}{8})-B(\frac{h}{4})}{15} & \overset{\rightarrow}{\searrow} & D(\frac{h}{8}) = \frac{64C(\frac{h}{8})-C(\frac{h}{4})}{63} & \searrow \\
\cdots & \cdots & \cdots & \cdots & \cdots & \cdots & \cdots & \cdots
\end{array}$$

gaining for each column two more powers of h in accuracy. If $A(h)$ and $A(\frac{h}{2})$ are two trapezoidal rule calculations, $B(\frac{h}{2})$ corresponds to Simpson's rule.

2 Named after Romberg's application of this extrapolation to improve the accuracy of the trapezoidal rule. While the procedure is robust, the convergence rate is quite slow, as increasing the order of accuracy by two requires a doubling of the number of function evaluations.

E.1.2 Using an Arbitrary Sequence of Step Sizes

Using steps $h_1, h_2, \ldots, h_k$, we can truncate (E.1) after the $h^{2(k-1)}$ term and then eliminate $c_1, c_2, \ldots, c_{k-1}$ by solving the linear system

$$\begin{bmatrix} 1 & h_1^2 & \cdots & h_1^{2(k-1)} \\ 1 & h_2^2 & \cdots & h_2^{2(k-1)} \\ \vdots & \vdots & & \vdots \\ 1 & h_k^2 & \cdots & h_k^{2(k-1)} \end{bmatrix} \begin{bmatrix} E \\ c_1 \\ \vdots \\ c_{k-1} \end{bmatrix} = \begin{bmatrix} A(h_1) \\ A(h_2) \\ \vdots \\ A(h_k) \end{bmatrix} \tag{E.4}$$

for E.[3] While this does not provide a sequence of approximations (as the Romberg table does, and which is useful for error estimates), there is now no need to decrease h at a high rate (which would rapidly increase computational costs). This extrapolation plays a key role in Section F.4.

E.2 Aitken Extrapolation

While this approach often is spectacularly effective, its theory is not straightforward (unsurprisingly, as it is a nonlinear procedure). If a sequence $\{x_n\} \to x$ converges *geometrically*, then

$$\begin{cases} x_n - x & = \lambda\,(x_{n-1} - x), \\ x_{n-1} - x & = \lambda\,(x_{n-2} - x), \end{cases} \tag{E.5}$$

and eliminating λ gives

$$x = x_n - \frac{(x_n - x_{n-1})^2}{x_n - 2x_{n-1} + x_{n-2}}. \tag{E.6}$$

The exact limit is in this case obtained from the values at any three successive values in the $\{x_n\}$ sequence.[4] Repeated use of (E.6) can give remarkably accurate results even when the geometric convergence assumption is grossly violated.

E.2.1 Case with Restarts

One simple application is for scalar root finding (then also known as Steffensen's method). Writing a scalar equation as $x = f(x)$, direct fixed point iteration

[3] The matrix is of Vandermonde type, and is thus nonsingular as long as no two steps sizes are the same, see Sections 1.2.1.1 and A.2.

[4] Aitken extrapolation is also known as Aitken's delta-squared process, since with $\Delta x_n = x_n - x_{n-1}$ (E.6) can be written as $x = x_n - \frac{(\Delta x_n)^2}{\Delta^2 x_n}$.

$x_{n+1} = f(x_n)$ requires for convergence (if starting close enough to x) that $|f'(x)| < 1$ and, if so, typically gives linear convergence. If starting close enough to x, but without any condition on $f'(x)$, the sequence

$$
\begin{array}{llllllll}
x_0 & x_1 & x_2 & \text{Iterate } x_{n+1} = f(x_n) & & & & \\
 & & \downarrow & \text{Aitken extrapolation} & & & & \\
 & & \widetilde{x}_0 & \widetilde{x}_1 & \widetilde{x}_2 & \text{Iterate } \widetilde{x}_{n+1} = f(\widetilde{x}_n) & & \\
 & & & & \downarrow & \text{Aitken extrapolation} & & \\
 & & & & \widetilde{\widetilde{x}}_0 & \widetilde{\widetilde{x}}_1 & \widetilde{\widetilde{x}}_2 & \text{Iterate} \ldots \\
 & & & & & & \downarrow & \longrightarrow
\end{array}
$$

generally gives quadratic convergence (number of correct digits double after each extrapolation). Compared to Newton's method, this procedure requires no derivative calculations, but does not generalize well to systems of equations.

E.2.2 Case without Restarts

In many applications, an infinite sequence is provided with no option to influence it by any kind of restarts. Such cases arise, for example, when approximating infinite sums.

Example E.2.1 Estimate the value of the infinite sum

$$s = \sum_{k=0}^{\infty} \frac{(-1)^k}{\log(k+2)} \approx 0.92429989722293885595957018136 \qquad \text{(E.7)}$$

by extrapolating a leading sequence of its partial sums.

Solution This is an extremely slowly converging sequence, with direct summation to reach 10^{-15} accuracy requiring in excess of $10^{434 \cdot 10^{12}}$ terms. The second column in Table E.1 gives the values of the first 17 partial sums. Whenever three consecutive values are available, we can perform one extrapolation according to (E.6). The third column in the table thus starts when the first three terms are available. Then extrapolating on these values starts after five original terms and so on. We see that using the first 17 partial sums of (E.7) has sufficed to reach machine precision. While extrapolation in general has a bad reputation for numerical sensitivity, this calculation was done in standard double precision, with essentially no digit lost. □

The effectiveness of Aitken extrapolation can be very problem dependent.[5] Some theory can be found in the references at the start of this Appendix. From

[5] For example, on $\sum_{k=1}^{\infty} \frac{1}{k^2} = \frac{\pi^2}{6}$, it does almost no good at all (possibly related to the fact that successive partial sums in this case do not bracket the limit). Numerical methods for infinite positive sums can benefit greatly when a corresponding integral is available, see Section 7.4.

Table E.1 *Repeated Aitken extrapolation applied to the first 17 partial sums of* (E.7).

Number of terms	Partial sum	Number of correct decimal digits ($-\log_{10}$ |error|)							
1	1.442695040888963	-	-	-	-	-	-	-	-
2	0.532455814262126	-	-	-	-	-	-	-	-
3	1.253803334706608	2.0	-	-	-	-	-	-	-
4	0.632468400146996	2.4	-	-	-	-	-	-	
5	1.190579026698243	2.7	3.7	-	-	-	-	-	-
6	0.676680684328492	2.9	4.2	-	-	-	-	-	-
7	1.157579031291480	3.1	4.6	5.5	-	-	-	-	-
8	0.702459417978061	3.3	4.9	6.1	-	-	-	-	-
9	1.136753899881313	3.4	5.2	6.5	7.5	-	-	-	-
10	0.719721508457067	3.5	5.4	6.9	8.1	-	-	-	-
11	1.122151112838911	3.7	5.7	7.3	8.6	9.7	-	-	-
12	0.732279867587631	3.8	5.9	7.6	9.0	10.2	-	-	-
13	1.111203049277582	3.9	6.0	7.9	9.4	10.7	12.0	-	-
14	0.741933676208727	3.9	6.2	8.1	9.7	11.1	12.5	-	-
15	1.102607436430968	4.0	6.4	8.3	10.1	11.6	13.0	15.0	-
16	0.749651312566207	4.1	6.5	8.6	10.4	12.0	13.4	14.8	-
17	1.095627568827401	4.2	6.6	8.8	10.6	12.3	13.9	16.0	15.0

a practical point of view, just extrapolating and then monitoring convergence rates in the table of successive extrapolations is often sufficient to gain some rough error estimates.

E.3 Taylor to Padé Conversion

Taylor expansions are best appreciated in the complex $z = x+iy$ plane. Expanded around $z = 0$, they take the form

$$T(z) = \sum_{k=0}^{\infty} c_k z^k \,, \tag{E.8}$$

convergent for $|z| < R$ and divergent for $|z| > R$ (where R may be 0 or ∞). If truncated to degree $m + n$, then

$$T_{m+n}(z) = \sum_{k=0}^{m+n} c_k z^k \tag{E.9}$$

contains equally many coefficients as

$$P_n^m(z) = \frac{\sum_{k=0}^{m} a_k z^k}{\sum_{k=0}^{n} b_k z^k}, \tag{E.10}$$

Algorithm E.1 Simple code for Taylor to Padé conversion.

```
function [a,b] = Pade(Na,Nb,c)

% Input parameters:
%    Na,Nb   Number of terms in the a and b-expansions, respectively
%        c   The c expansion - column vector of length N.
%            It should hold that Na+Nb = N+1
% Output parameters:
%      a,b   Column vectors of lengths Na,Nb, respectively, with b(1) = 1.

N = length(c);
A = [toeplitz([1,zeros(1,N-1)]',[1,zeros(1,Na-1)]),...
               -toeplitz(c,[c(1),zeros(1,Nb-1)])];
rhs = -A(:,Na+1);  A(:,Na+1) = [];     ab = A\rhs;   % Form and solve the
                                                     % linear system
a = ab(1:Na);      b = [1;ab(Na+1:N)];          % Return the coefficient
                                                % vectors a and b
end
```

where we assume $b_0 = 1$, and do not count this coefficient. Requiring $T_{m+n}(z)$ and $P_n^m(z)$ to match as well as possible around $z = 0$, that is,

$$T_{m+n}(z) - P_n^m(z) = O(z^{m+n+1}) \tag{E.11}$$

allows the (truncated) sequences $\{a_k\}$, $\{b_k\}$ to be converted to $\{c_k\}$, and vice versa.[6] The conversion $\{a_k\}, \{b_k\} \to \{c_k\}$ becomes explicit after multiplying (E.11) with the denominator in (E.10) and equating coefficients. The reverse conversion $\{c_k\} \to \{a_k, b_k\}$ also requires the solution of a linear system. The straightforward code for this that is shown in Algorithm E.1 does not include any special treatment of possible singular or near-singular cases (discussed in detail in Gonnet et al., 2013).

The theory for Padé expansions (such as convergence properties) is complicated, and we refer (also for its connection to continued fractions) to Bender and Orszag (1978); Brezinski and Zaglia (1991). Extensive historical perspectives are provided in Brezinski (1991); Brezinski and Zaglia (2020). While numerical *analytic continuation* is a notoriously ill-conditioned problem (Trefethen, 2020, 2023), the rational form of $P_n^m(z)$ provides implicitly some highly regularizing constraints on how fast the approximation can grow (if at all) with $|z|$.[7] From this it follows that $T_{m+n}(z) \to P_n^m(z)$ conversions can provide excellent

[6] The description here is slightly simplified. In exceptional cases, the exponent in the right-hand side of (E.11) may be different.

[7] When given a truncated Taylor expansion of degree d, one typically chooses $m = n = d/2$ for d even and $m + 1 = n = (d + 1)/2$ for d odd.

extrapolations of the function $T(z)$ also well outside the region of convergence $|z| < R$ for its Taylor expansion, as illustrated in the following example.[8]

Example E.3.1 Given only a number of leading coefficients in the Taylor expansion for

$$f(z) = \frac{\log(1+z)}{z} = 1 - \frac{1}{2}z + \frac{1}{3}z^2 - \frac{1}{4}z^3 + - \cdots ,$$

approximate $f(2)$ (for which the analytical value is $\frac{1}{2}\log 3 \approx 0.549306144$).

Solution Figure E.1(a) shows the analytically correct $\operatorname{Re} f(z)$ over the complex plane region $[-2.5, 2.5] \times [-2.5, 2.5]$. There is a singularity at $z = -1$, with a branch cut to its left. Part (b) shows similarly the result of its Taylor expansion truncated to degree 20. Due to the singularity at $z = -1$, including increasingly many terms improves the accuracy for $|z| < 1$ and makes the divergence worse for $|z| > 1$.[9] Converting this $T_{20}(z)$ expansion to $P_{10}^{10}(z)$ rational form gives (again displaying the real part) the result shown in Figure E.2(a). As this is a rational (single-valued) analytic function, it approximates the branch cut by a sequence of poles.[10] However, away from the branch cut along $z \leq -1$, the accuracy is very good. As seen in part (b), the accuracy (shown at $z = 2$) improves exponentially fast with increasing values for $m = n$, while the Taylor truncated representation of degree $m + n$ then diverges exponentially fast. □

A more extreme example is provided by the Stieltjes function

$$f(z) = \int_0^\infty \frac{e^{-t}}{1+zt}\,dt = \sum_{k=0}^{\infty} (-1)^k k!\, z^k .$$

Although the radius of convergence in this case is $R = 0$ (due to a branch point at $z = 0$), conversion to Padé rational form allows also in this case numerical evaluation (based only on the Taylor coefficients $c_k = (-1)^k k!$) to any precision away from its branch cut (now along $z \leq 0$). It is a commonly held misconception that the radius of convergence in some fundamental way limits the region in the complex plane over which a Taylor series can provide accurate information.

An important application of Padé extrapolation arises in the context of (usually divergent) asymptotic expansions, where maybe only a few terms can be

[8] A very different application of $T_{m+n}(z) \to P_n^m(z)$ conversions is described in Section 1.2.1.5.

[9] Part (b) of the figure also illustrates Jentzsch's theorem: If a power series has a finite radius of convergence, the zeros of partial sums will be dense around the boundary of convergence.

[10] In the case of a single branch point, Padé expansions commonly place an alternating sequence of poles and zeros in its "shadow" (as seen from the expansion center) (Stahl, 1997).

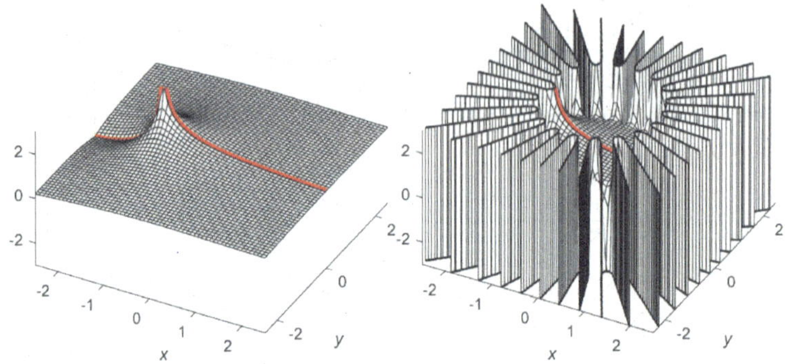

(a) True function Re $f(z)$. (b) Real part of its truncated expansion $T_{20}(z)$.

Figure E.1 The function $f(z) = \frac{\log(1+z)}{z}$ and the sum of its Taylor expansion up through degree 20. Values along the real axis are highlighted in red.

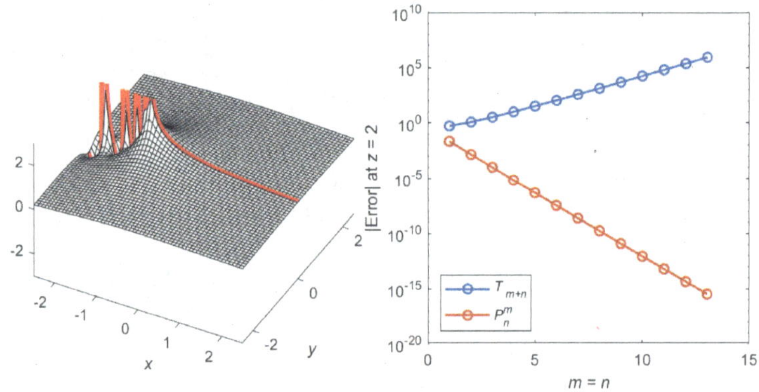

(a) Real part of the Padé approximation $P_{10}^{10}(z)$. (b) Error in Taylor and Padé approximations.

Figure E.2 (a) The Padé approximation $P_{10}^{10}(z)$ obtained from the $T_{20}(z)$ truncated Taylor sum shown in Figure E.1(b). (b) The errors in the Taylor approximations $T_{2n}(2)$ and matching Padé approximations $P_n^n(2)$ for $n = 1, 2, \ldots, 13$. As the plot is log–linear, the trends are seen to be exponential divergence versus exponential convergence.

obtained with reasonable effort, and most information possible then needs to be extracted from these.

Example E.3.2 Approximate the same alternating infinite sum

$$s = \sum_{k=0}^{\infty} \frac{(-1)^k}{\log(k+2)}$$

as in Example E.2.1 by using Padé conversion.

Solution In order to apply the Padé approach, we consider the Taylor series $T(z) = \sum_{k=0}^{\infty} \frac{z^k}{\log(k+2)}$ and choose $m = 8$, $n = 8$ (i.e., as in Example E.2.1 using only the first 17 terms of the series) to obtain $P_8^8(-1)$ with an error of $1.5 \cdot 10^{-13}$, that is, slightly less accuracy than with the Aitken extrapolation. However, as usual when approximating infinite sums, evaluating some leading terms explicitly before applying an extrapolation technique greatly improves the accuracy (see the curves in Figure 7.2). In the present example, summing just four leading terms and then approximating $T(z) = \sum_{k=4}^{\infty} \frac{z^k}{\log(k+2)}$ with its first 17 terms suffice to bring the error after evaluating $P_8^8(-1)$ to machine precision $3.3 \cdot 10^{-16}$. □

Appendix F

Trade-offs between Accuracy Orders and Other Approximation Features

The theme of this book is high-accuracy FD methods. Only for pseudospectral methods (Chapter 2) was the accuracy increased to its infinite order limit. Some compromises between order of accuracy and other features (such as handling of boundaries, numerical stability, etc.) are usually needed. This appendix focuses on cases where high orders of accuracy are readily achievable, but some of these orders can favorably be exchanged for the introduction of free parameters which then can be utilized for alternative purposes.

Lowering the order of accuracy in otherwise very high order methods is not the only way to enhance the handling of specific solution properties. "Nonstandard finite differences" (NSFD) (Mickens, 2002, 2020) aim at incorporating certain features (such as conservation of energy, monotonicity, positivity, etc.), but are generally limited to FD approximations of relatively low orders of accuracy. The related concept of "Mimetic finite differences" (MFD), also designed to mimic select solution features, is surveyed in Lipnikov et al. (2014).

F.1 Trade-off between Accuracy Order and Wave Number Range

Figure 2.1 displayed the factor a Fourier mode $e^{i\,\omega x}$ gets multiplied with when different centered FD approximations for d/dx are applied to it. In the context of solving the simple convection PDE (4.8) $\frac{\partial u}{\partial t} + \frac{\partial u}{\partial x} = 0$, we can read off the error in propagation speed for mode ω as the vertical deviation from each curve to the ideal diagonal line, and then divided by ω. Especially for high-order approximations, the extreme accuracy near $\omega = 0$ can in some applications be advantageously traded against distributing the error more uniformly across a

Table F.1 *The weights in standard FD approximations for the first derivative vs. weights optimized for wider bandwidths.*

Stencil width	Increasing accuracy orders Right half of Table 1.1				Second-order accuracy Weights as given in Table 2 of Kumari et al. (2019)			
3	$\frac{1}{2}$							
5	$\frac{2}{3}$	$-\frac{1}{12}$			0.9415	−0.2208		
7	$\frac{3}{4}$	$-\frac{3}{20}$	$\frac{1}{60}$		0.9116	−0.3730	0.1114	
9	$\frac{4}{5}$	$-\frac{1}{5}$	$\frac{4}{105}$	$-\frac{1}{280}$	0.9393	−0.3764	0.1821	−0.0582

wider part of the frequency range $-\frac{\pi}{h} < \omega < \frac{\pi}{h}$.[1] As an illustration, the left part of Table F.1 copies the right half of Table 1.1, showing centered FD weights for the first derivative, approximated to increasing orders of accuracy. Instead using the weights shown in the right part (here rounded to four decimal places from Table 2 of Kumari et al. (2019)) keeps the accuracy at second order but broadens the range of frequencies that are reasonably well approximated, as seen in Figure F.1. In applications where the frequency band of interest is well defined (as for waves initiated by a known oscillator), high accuracies can be realized with quite narrow FD stencils. This idea has been extensively developed since it was first proposed in the early 1970s to also include implicit (compact) FD approximations (Deshpande et al., 2021), grid staggering (Section 4.1.4), and combined space-time discretization (Section 4.1.2). A small dissipative term can be added to damp out select modes as they lose their accuracy. Schemes of this type are widely used in signal processing applications (such as for seismic modeling Holberg, 1987). Surveys include Kumari et al. (2019); Sagaut et al. (2023); Vichnevetsky (1987); Zingg (2000).

F.2 Enhanced Gregory Quadrature

For numerical quadrature based only on values of a smooth function at equispaced nodes spanning the integration interval, the trapezoidal rule (TR) is spectrally accurate within domain interiors, and the Gregory approach (described in Section 7.3) is effective for end corrections. Sections F.2.1 and F.2.2 describe accuracy–order trade-off opportunities.

[1] Somewhat conceptually similar to the error redistribution over an interval when changing from a Taylor to a Chebyshev expansion.

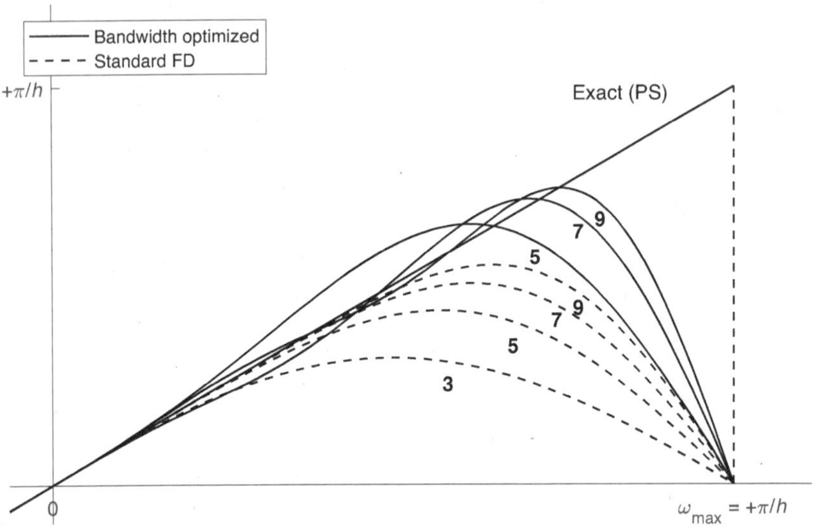

Figure F.1 The factor the Fourier mode $e^{i\,\omega x}$ gets multiplied by (omitting a factor of i) when differentiated by the different FD approximations given in Table F.1. Solid curves: Second-order accurate and then optimized for bandwidth. Dashed curves: Standard FD: Optimized for accuracy order (degree of fit at the origin). The numbers indicate the stencil widths.

F.2.1 Case When the Integration Interval Starts and Ends on Grid Points

The red circles in Figure F.2 show the Gregory weights in the cases of accuracy orders $p = 4, 10, 14$, followed by light blue circles indicating weights that are exactly one.[2] A few observations:

i. The top subplot ($p = 4$) compares the Gregory weights to Simpson's formula, which for increasing orders generalizes to the Newton–Cotes formulas. These latter are highly oscillatory throughout the entire domain interior (and are nowadays rarely used beyond the Simpson case).
ii. For order p, the Gregory weights have $p - 1$ nontrivial entries. As p is increased, their magnitudes grow rapidly, causing at $p = 10$ the first instance of a negative weight.

Since Gregory formulas of arbitrarily high orders of accuracy are trivially obtained, this naturally suggests trading some of these orders for the different

[2] If the node spacing is h rather then one, all weights need to be divided by h.

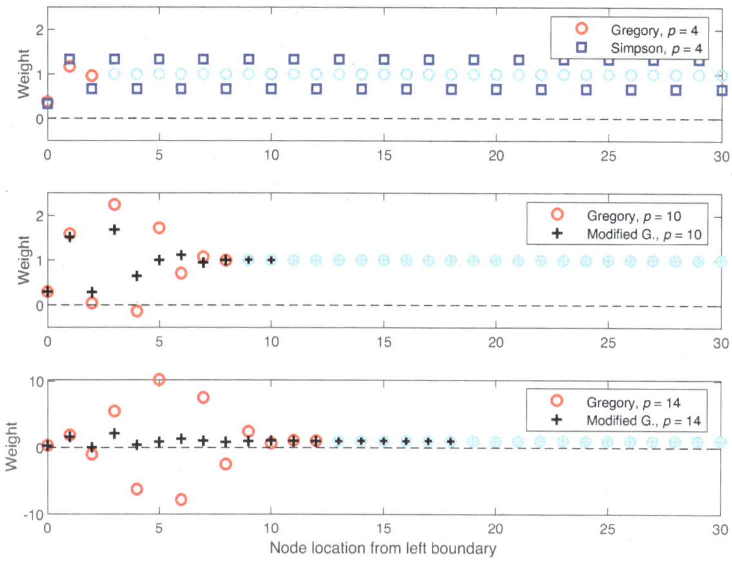

Figure F.2 Comparison of weight sets for TR corrections (shown for the left boundary) in cases of accuracy orders $p = 4, 10, 14$. Note the different vertical scale in the last subplot. Light blue markers show weights equal to 1.

goal of ensuring that all weights are nonnegative. The last two subplots of Figure F.2 illustrate this concept, with black crosses changing to light blue crosses for weights identically 1. We refer to Fornberg (2021e) and Fornberg and Lawrence (2023) for algorithmic details (and MATLAB codes).

Gregory corrections (to which 1 should be added to obtain the weights) were shown in Table 7.1, with the bottom row corresponding to the $p = 10$. The case "Modified G., $p = 10$" in Figure F.2 shows the alternative (also 10th-order accurate) set of weight corrections

$$\frac{1}{504}\left\{\frac{-22763}{64}, \frac{59501}{225}, \frac{-64849}{180}, \frac{11027}{32}, \frac{-40069}{225}, \frac{6071}{7200}, \frac{45847}{800}, \frac{-40171}{1440}, \frac{-289}{2880}, \frac{2917}{800}, \frac{-1957}{2400}\right\}. \tag{F.1}$$

This uses two more nontrivial weights than Gregory of order $p = 10$, but does not lead to any negative weights. For still higher orders of accuracy, the benefits of this weight reduction strategy are even more striking, as seen in the third subplot of Figure F.2. Beyond (F.1), these improved higher-order weight sets are not easily expressed as rational numbers but are readily computed numerically.

Figure 8.3(a) showed the test function $f(x) = \cos(20\sqrt{x})$ over $[0, 1]$, and Figure F.3 shows the error obtained with four of the quadrature methods just described, when using step sizes from $h = 10^{-1}$ down to $h = 10^{-3}$ (with all cal-

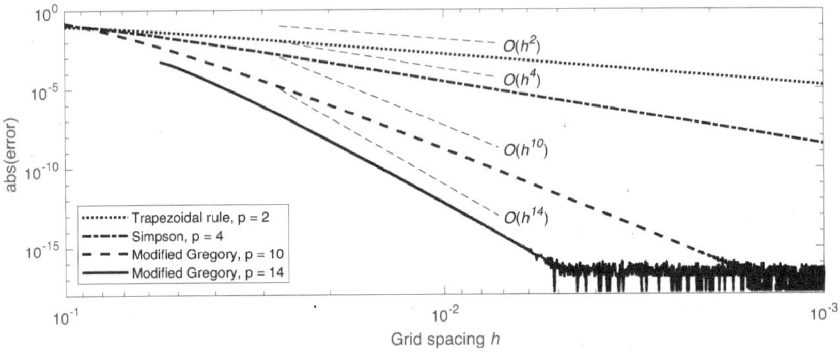

Figure F.3 Convergence rates for four quadrature methods, when applied to the test function in Figure 8.3(a).

culations in double precision[3]). Convergence occurs at the rates to be expected from the formal orders of the methods. As their cost is the same (to leading order, for same h), the benefit of using high orders of accuracy is obvious.

F.2.2 Case When the Integration Interval Does Not Start and/or End at a Grid Point

As we have noted repeatedly, data in physical applications is often available only at equispaced locations. However, the function to be integrated may be discontinuous at some known (nongrid-point) location, with the equispaced data on the two sides unrelated. Alternatively, integration may be desired only between nongrid-point locations, with data unavailable outside the integration interval. A variation on the theme of Section F.2.1 above is again available, as described in Fornberg and Lawrence (2023),[4] reaching 10th-order accuracy with, as before, all weights nonnegative. Figure F.4(a) illustrates the test problem

$$\int_0^1 f(x)dx \quad \text{with} \quad f(x) = \begin{cases} e^{-3x}\sin(20x), & 0 \le x < 1/\sqrt{2}, \\ -\frac{2}{5}\cos(10x), & 1/\sqrt{2} \le x \le 1, \end{cases} \tag{F.2}$$

featuring a discontinuity at a location that does not under any level of refinement with equispaced nodes over $[0, 1]$ align with a data point. While this

[3] As the calculations do not involve any division by h, rounding errors do not cause any overall error growth for very small h (as was seen in Figure 1.2). Compared to the results of direct Gregory calculation shown in Figure 8.4(a), the present schemes are much more robust against noise in function values as all weights now are nonnegative (and much smaller in magnitude).

[4] with a brief MATLAB code included

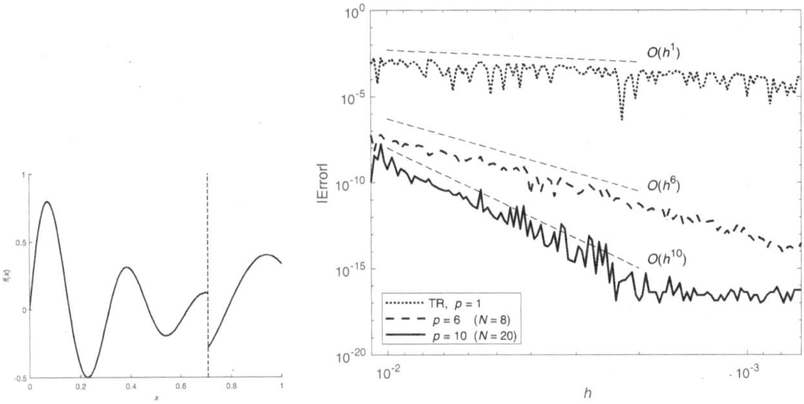

(a) Test function (b) Convergence rates

Figure F.4 Test function and enhanced Gregory quadrature convergence rates for the discontinuous function test case described in Section F.2 (using end corrections extending over N nodes).

causes some jitters in the quadrature results shown in part (b) of the figure,[5] convergence orders become nevertheless as theoretically predicted.

F.3 RBF–FD at Boundaries

The concept is somewhat similar to the enhanced Gregory method in the previous Section F.2, but with the convenient difference that the edge corrections occur naturally, requiring no special attention. We recall from Section 5.4.1 that the order of accuracy of PHS+poly approximations are determined by the degree of the supplementary polynomials. For regular one-sided FD approximations, high degrees will at/near boundaries cause a severe Runge phenomenon. This was seen, for example, in Table 1.3 (approximating the first derivative at a boundary node), with the magnitude of these weights for accuracy orders 1–7 displayed in Figure F.5(a). Part (b) of this figure shows similarly the weights when the stencil is extended still further into the domain, but with approximations using PHS+poly rather than just poly, with the polynomial degree locked in at 7, ensuring accuracy order 7. The front row in the figure's Part (b) matches the back row in Part (a). Extending the stencil size is seen to make the weights

[5] having used as input data only the equispaced function values and knowledge of the discontinuity location

near the boundary revert toward those of low-order FD approximations (in spite of the higher accuracy order being preserved).[6]

As described further in Bayona et al. (2019) (with additional theoretical perspectives in Bayona (2019b)), corresponding benefits hold also in higher dimensions, reducing or eliminating the need for "ghost points" external to a domain.[7,8]

Example F.3.1 With the domain and node set shown in Figure F.6, approximate the eigenvalues of the Laplace problem $\left(\frac{\partial^2}{\partial x^2}+\frac{\partial^2}{\partial y^2}\right)u=\lambda u$ with zero Dirichlet boundary conditions, using RBF–FD PHS+poly discretization at all interior nodes.

Solution Choosing $\phi(r)=r^3$, supplementary polynomials of degree 4, and stencils of sizes $n=16,21,31$ give the results shown in Figure F.7. Mathematically, all eigenvalues should be real and negative. While the eigenvalues for the smoothest eigenfunctions are quite accurate in all cases, the eigenvalues for higher eigenfunctions display a large (and unphysical) scatter in cases of smaller stencils, mostly related to edge effects. Simply increasing the stencil size, keeping the accuracy order (polynomial degree) fixed, greatly reduce (or overcome) these errors. □

F.4 Parallel-in-Time ODE Solver for Wave Equations

Most modern computers have multiple cores, which can be run in parallel. In many cases, it is straightforward to partition a larger task into independent subtasks, for increased throughput speed. However, time stepping ODEs is an intrinsically sequential task. Nevertheless, some limited opportunities for parallel processing are available.[9] In the context of time-stepping wave equations, the GBS scheme (Section 3.1.4) provides the opportunity to effectively use up to around eight cores in parallel.[10] The following is a brief summary the concept of this approach, with more information available in Ellison and Fornberg (2021).

The extrapolation methods we consider here combine some number M of

[6] This result is heuristically plausible when comparing (5.14) and (D.12), showing that solving for RBF-FD weights corresponds to a minimization of the norm of the weights vector $\underline{w}$.

[7] Similarly to the present Figure F.5, Figure 2 in Bayona et al. (2019) illustrates the case of approximating the second derivative at a boundary point when using one ghost point outside the domain.

[8] The idea for the enhanced Gregory approach described in Section F.2 originated from the (surprise) observation that numerical quadrature based on integrating PHS+poly approximations naturally bypasses the Runge phenomenon (Reeger and Fornberg, 2018).

[9] Developments over the previous 50 years were surveyed in Gander (2015).

[10] The Parareal strategy is in this context ineffective (Ruprecht, 2018).

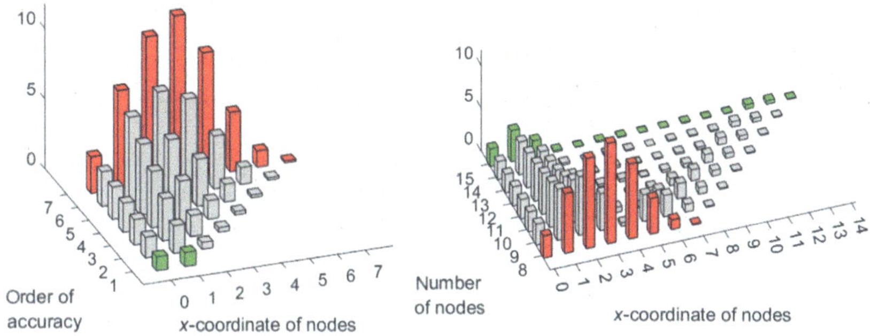

(a) Lines 1–7 in Table 1.3 (b) Weights when stencil sizes increased

Figure F.5 Magnitudes of weights (a) Regular FD approximations for the first derivative at the left boundary node, up through order 7 (which requires 8 nodes), and (b) Polynomial degree (and order of accuracy) fixed at 7, but with increasing stencil sizes and including a RBF of type $\phi(r) = r^3$ at each node.

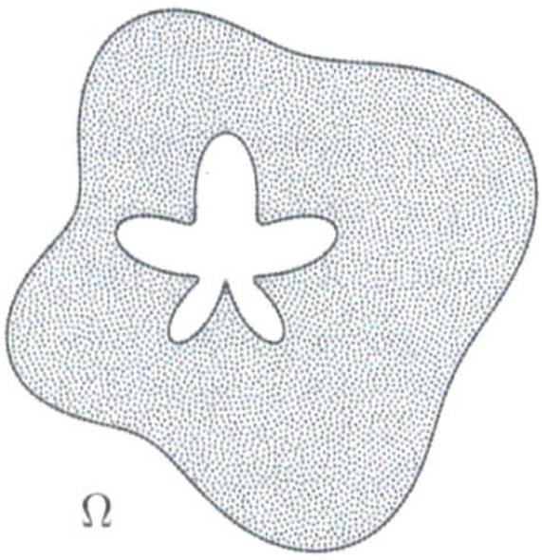

Figure F.6 Domain and node set considered in Example F.3.1. Figure reproduced from Bayona et al. (2017).

GBS-generated sequences with different N-values (number of internal steps, with matching internal step lengths k).[11] The extrapolations can then produce a result with leading error term $O(k^{2M})$. When time-stepping wave-type PDEs, the critical factor, besides order of accuracy, is that the method's stability domain covers as long stretch as possible of the imaginary axis in the complex ξ-plane. For each method, this interval can be denoted by $i\,[-\mathrm{ISB_n}, \mathrm{ISB_n}]$, with ISB abbreviating "Imaginary Stability Boundary" and the subscript n abbreviating "normalized," indicating that, for fair comparison between methods, the

[11] Stability diagrams for various M-values and N-sequences are shown in Fornberg et al. (2007).

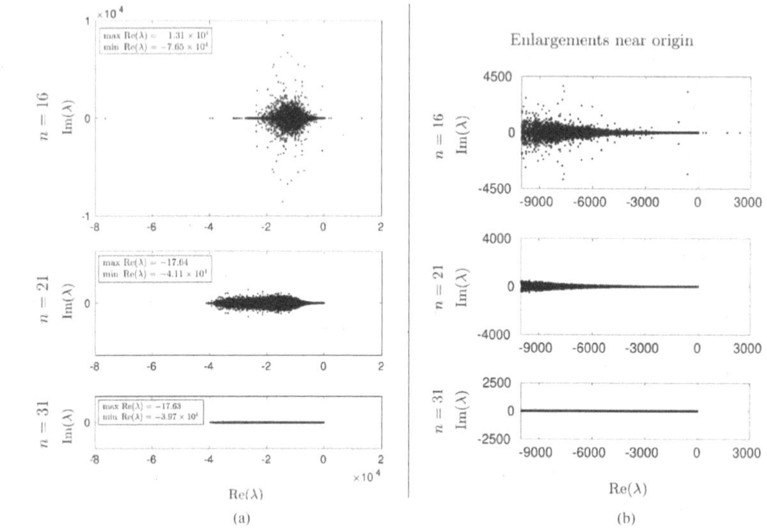

Figure F.7 Computed eigenvalues in Example F.3.1. The three subplots in (a) have the same horizontal and vertical scales. Figure reproduced from Bayona et al. (2019).

actual ISB has been divided by the number of stages (sequential function evaluations) the method requires.[12] The methods developed in Ellison and Fornberg (2021) amount to giving up several of the $2M$ orders of accuracy in exchange for obtaining a corresponding number of free parameters, which then can be utilized to maximize the $\mathrm{ISB_n}$.[13]

Figure F.8 summarizes the $\mathrm{ISB_n}$ values for a number of GBS schemes, of accuracy orders 4, 8, 12, 16 optimized as described earlier, relative to standard RK4 and RK8 schemes. The $\mathrm{ISB_n}$ values are inversely proportional to required computer wall-clock time. For example, RK4 (which can only utilize one core) and GBS12 using eight cores have roughly the same $\mathrm{ISB_n}$ value and will run in roughly the same wall clock time for matching step sizes. However, with the large difference in accuracy orders, the GBS12 scheme gets a vast performance advantage.[14] Figure F.9 shows the eight-core GBS12 scheme's stability domain in the complex ξ-plane, with its $\mathrm{ISB_n} \approx 0.7116$ close to the value $\sqrt{2}/2 \approx 0.7071$ for RK4.

[12] The upper $\mathrm{ISB_n}$ limit for any explicit method is 1; for fourth-order accurate explicit methods (not utilizing "background cores"), RK4 is optimal, with $\mathrm{ISB_n} = (2\sqrt{2})/4 \approx 0.7071$ (Kinnmark and Gray, 1984).

[13] Leading to a convex optimization problem studied earlier in a RK context (Ketcheson and Ahmadia, 2012).

[14] See Figure 3.2, which shows typical results for when ODE methods of different accuracy orders are compared (for matching step sizes k).

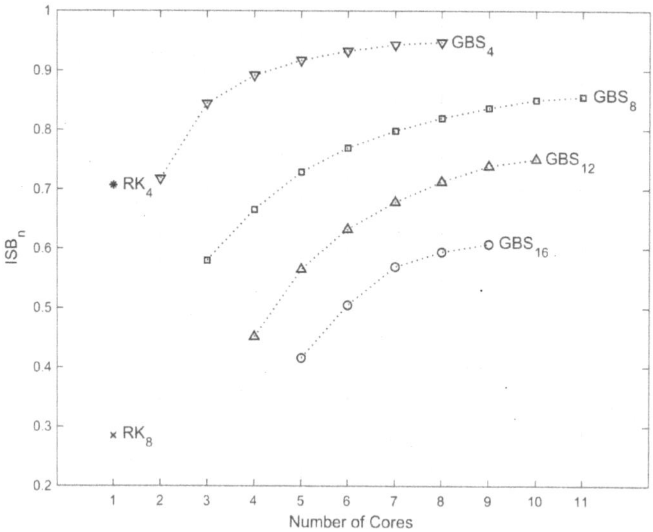

Figure F.8 Comparisons between the ISB_n values for RK4 and RK8 vs. ISB_n-optimized GBS schemes for different counts of available cores. Figure reproduced from Ellison and Fornberg (2021).

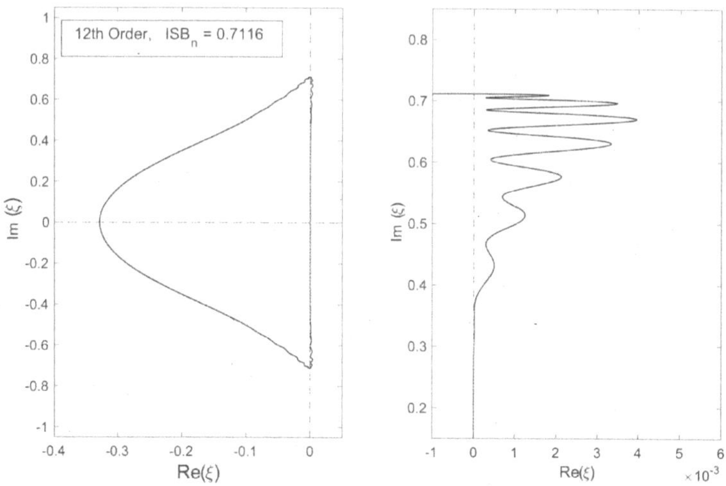

Figure F.9 The stability domain (normalized by n) for the ISB_n-optimized eight-core GBS_{12} scheme. Figure reproduced from Ellison and Fornberg (2021).

Appendix G
Node Sets for FD and RBF-FD-based PDE Discretizations

Figure G.1 illustrates the concepts for the types of node sets that are most relevant in our present context. The two main sections below focus on the "Grids" and "Mesh-free nodes" cases, respectively. Grid-based discretizations were traditionally the only option in FD contexts. In contrast, mesh-free[1] methods do not require any direct node-to-node connections. The nodes can be placed irregularly, thereby offering simple opportunities for gradual local refinement in critical spatial areas. Finite-element methods are in this sense not meshfree but rather based on irregular meshes, as they require specific node-to-node connections.

G.1 Grids

G.1.1 Cartesian Grids

These node sets have historically been the almost exclusive choice in FD contexts. The nodes can have variable spacing in each of the spatial directions. Fundamental strengths of Cartesian node layouts include coding simplicity and trivial generalizations to any number of dimensions. With equispaced nodes, symmetries can improve accuracy, computational stability, and simplify theoretical analysis.

G.1.2 Sparse Grids

The goal with these is to reduce the "curse of dimensionality" by omitting most of the internal node points in Cartesian grids, as schematically illustrated for 2-D in Figure G.2. The concept becomes increasingly effective as the number

[1] Also described as meshless.

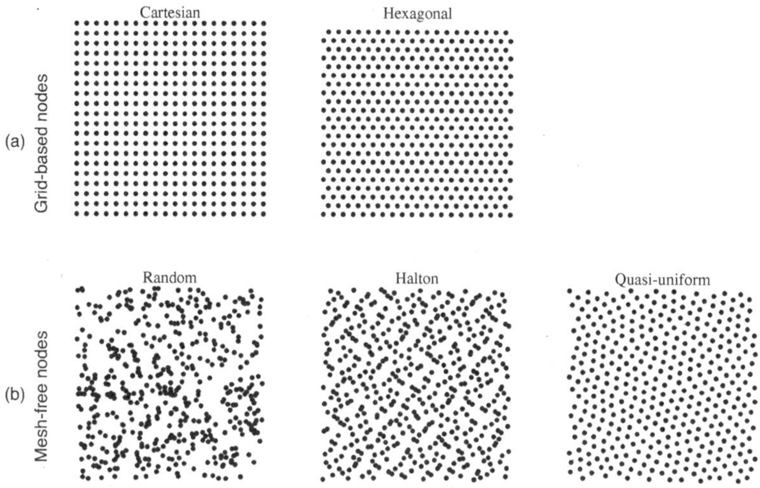

Figure G.1 Examples of different grid-based and mesh-free constant density function node sets in 2-D.

Figure G.2 Schematic illustration of the sparse grid concept in 2-D, for the first few levels of node refinement.

of dimensions d increases. It was introduced in Smolyak (1963) and is further described for PDEs and quadrature in Bungartz and Griebel (2004); Gerstner and Griebel (1998), respectively. It falls outside our main focus, as the resulting accuracy levels inevitably become quite low.

G.1.3 Hexagonal Grids

Before the arrival of finite elements and RBF-based discretizations, hexagonal grids[2] formed the main alternative to Cartesian grids. These may in some cases have a slight advantage in being somewhat more isotropic (in 2-D, with three rather than two primary grid directions) (Zingg and Lomax, 1993). However, implementation of boundary conditions becomes more involved, and extensions

[2] Considered already before the era of computers (Southwelll, 1945).

to above 2-D are impractical. At present, the simplest way to use hexagonal node sets is to simply regard them as a special case of quasi-uniform ones, with the benefit that weights in RBF–FD stencils can be extensively reused rather than having to be calculated separately at all locations. In 2-D, hexagonal node layouts may come close to being optimal in terms of accuracy, as long as no local refinement or other geometric flexibility is called for.

G.2 Mesh-Free Nodes

The three sets shown in the subplots in Figure G.1(b) have exactly the same number of nodes. They differ visually, but more importantly in how well they can support (generalized) FD discretizations (in particular RBF and RBF–FD methods). The random and quasi-uniform node sets generalize easily to spatially variable node densities, as needed in cases of local refinement in critical areas.

G.2.1 Fully Random Nodes

It is in the nature of complete randomness that various artifacts will occur, for example, tight node clusters, voids, random alignments along short curve segments, and so on (all visible in the first subplot of Figure G.1(b)). All these features are severely damaging in the context of approximating derivatives and solving PDEs.

G.2.2 Halton Nodes

Different types of low-discrepancy (quasi-Monte Carlo) node sets have been designed, falling somewhere in-between being fully random and quasi-uniform (as described later), with Halton nodes an important special case. These are readily generated in any number of dimensions.[3] The algorithm behind them includes a mathematical principle that prevents any two nodes from getting extremely close to each other. Practical RBF–FD experiences show that, even so, the remaining irregularities are too large for these node sets to be competitive in PDE contexts. There are, however, other contexts where Halton (and similar) node sets can be attractive. One such case is Monte Carlo methods for quadrature in high dimensions.[4] As a (low-dimensional) illustration, Figure G.3 compares

[3] Halton sequences are available in MATLAB's Statistics and Machine Learning toolbox and can alternatively be coded in less than a dozen lines, see Fornberg and Flyer (2015b), Section C.1.

[4] Different aspects of Monte Carlo type methods are surveyed in Caflisch (1998); Giles (2015); Hung (2023).

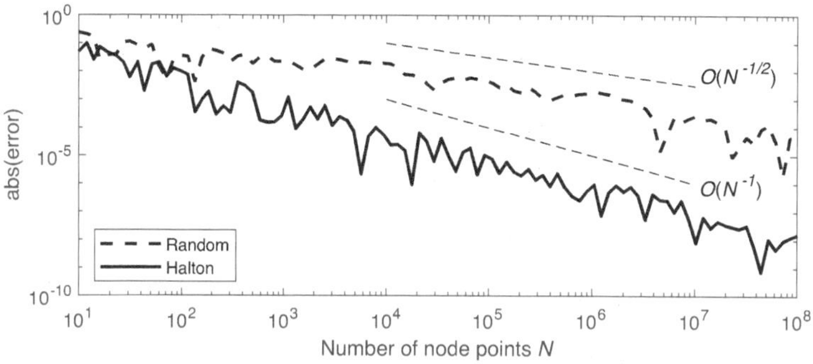

Figure G.3 Convergence rates for N-node Monte Carlo approximations of $\int_0^1 \cos(20\sqrt{x})\, dx$, when using random and Halton nodes, respectively.

the convergence rates when approximating $\int_0^1 \cos(20\sqrt{x})\, dx$ by averaging the value of the integrand at N random and N Halton points, respectively, across $[0, 1]$ (with the integrand shown in Figure 8.3(a)). While neither of these two versions compares well against grid-based quadrature (see Figures 8.4(a) and F.3 for the same test problem), the "curse of dimensionality" can still favor Monte Carlo approaches in high dimensions (as arise in contexts, such as quantum mechanics, mathematical finance, etc.).

G.2.3 Quasi-uniform Nodes

Numerous algorithms are available for creating quasi-uniform node sets, with prescribed spatially variable node densities. These either create nodes only, or additionally connect these into elements with favorable aspect ratios (for purposes such as computer graphics, finite elements, etc.), for example, Persson and Strang (2004). The step of creating elements can be costly (especially above 2-D) and is unnecessary in RBF–FD contexts. Algorithms focused on RBF–FD include Fornberg and Flyer (2015a); Slak and Kosec (2019); van der Sande and Fornberg (2021).[5] These may first distribute nodes across a geometrically simple domain, with application-specific boundaries (and internal interfaces) then incorporated in a separate step, as illustrated schematically in Figure G.4.

One approach for creating quasi-uniform node sets is based on simulating freely moving repelling electrical charges (of spatially variable strengths).[6] However (as discussed in Fornberg and Flyer, 2015a), this can become extremely costly if the initial guess of the node distribution is far away from its

[5] The latter two references describe quasi-uniform node set generation also in 3-D.
[6] Best results are often obtained if the repelling force is exaggerated for short distances.

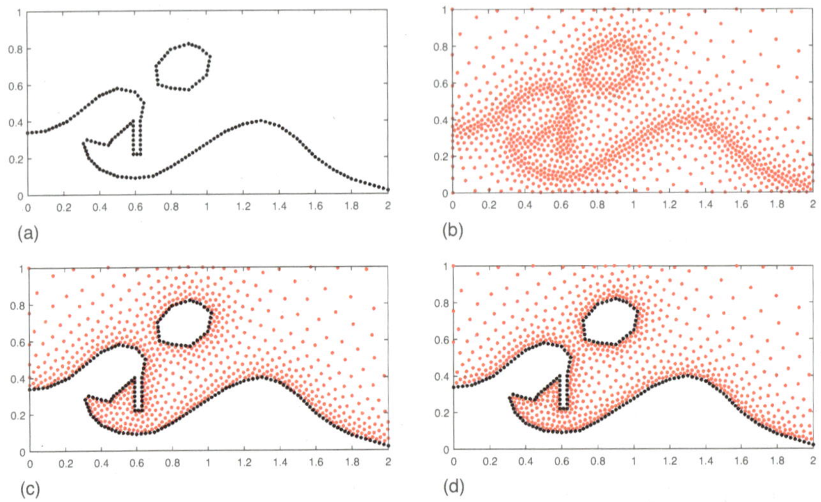

Figure G.4 Schematic illustration of a sequence of steps for creating quasi-uniform nodes to model waves near to a harbor. (a) Nodes along shore lines. (b) Quasi-uniform nodes across domain (computed using the brief MATLAB code in Fornberg and Flyer (2015a) with the node density set as a function of the distance to the nearest shore node). (c) Superposition of (a) and (b) with nodes on land removed. (d) Nodes after adjustment of nodes near the boundary relative to those on the boundary (dependent on the application, this last step may be unnecessary, such as when solving elliptic PDEs; see Nielsen and Fornberg, 2024).

eventual equilibrium,[7] but is highly efficient in the case when only very small position adjustments are needed (e.g., for adjusting nodes near boundaries, as in the last step shown in Figure G.4). A simple implementation strategy for such adjustments is as follows: For each node, find some number of its nearest neighbors (by a kd-tree search). One can then carry out a few iterations of determining the direction (ignoring strength) of the electrostatic force from these nearby nodes, followed by moving the node in this direction some gradually decreasing fraction of the local average node separation distance. If needed, this can be repeated (starting with a new kd-tree search). Figures G.5 and G.6 show examples of node sets obtained in this way for the purpose of modeling impacts of storm surges and tsunamis.

[7] Nodes will then move according to principles of diffusion rather than far more effective convection.

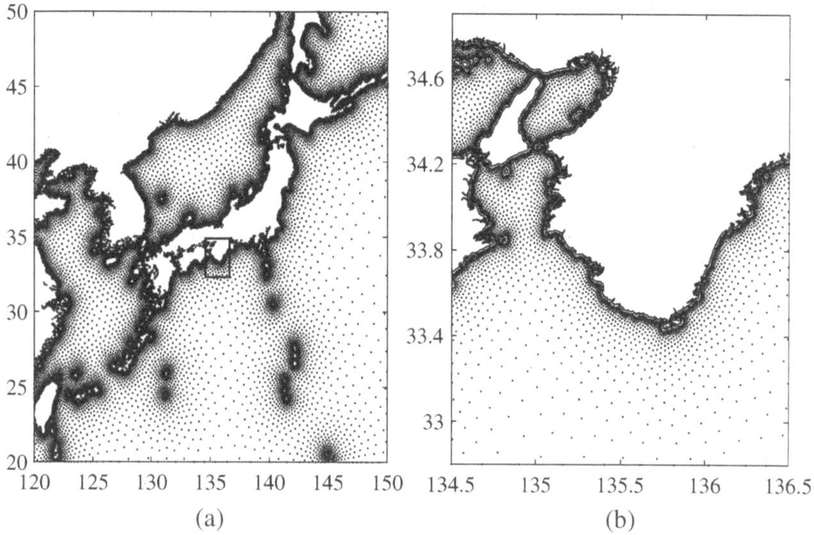

Figure G.5 Coarse node set surrounding Japan (a), with the region around Osaka Bay marked with a rectangle refined (b). The smallest node separations (closest to the coastlines) are in these plots 7.4 km and 0.64 km, respectively. The axes show latitude and longitude. Illustration courtesy of Adrean Webb, Tokyo Institute of Technology.

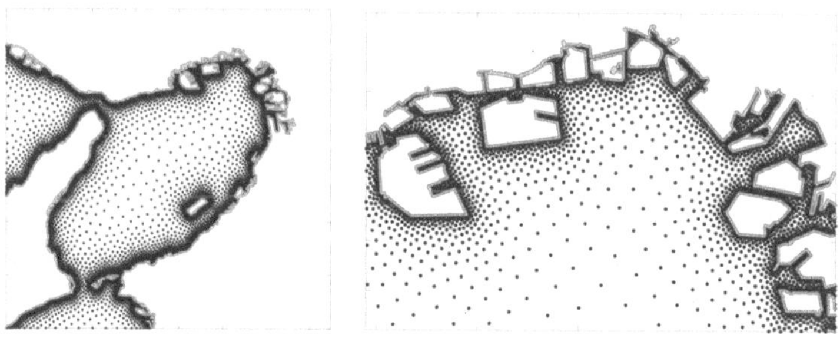

Figure G.6 Further successive node refinements of Osaka Bay and its harbor area. Illustration courtesy of Adrean Webb, Tokyo Institute of Technology.

G.2.4 Node Subsampling

Coarser node sets that are *subsampled* from finer ones (i.e., with all nodes on the coarser set present also on the finer set) are very suitable for multigrid applications (see Section 4.3.5), as this simplifies and improves the accuracy of data transfers between the levels. For Cartesian and Hexagonal node sets,

Figure G.7 Illustration of subsampling of a highly spatially variable density quasi-uniform node set. Figure adapted from Lawrence et al. (2024).

subsampling is readily achieved by using grid spacings that are increased by a small integer factor (typically two). Both Random and Halton node sets are built up sequentially by successive insertion of nodes, meaning that subsampling can be realized by recording earlier stages of the process. However, these latter two types of node sets are too irregular to be suitable for multigrid/multilevel applications.

With quasi-uniform node sets of spatially variable density being excellent for RBF–FD (and other) mesh-free applications, the question arises how these can be subsampled in some automated way that preserves key node set quality features. As when generating quasi-uniform node sets, it turns out that *moving front algorithms* are particularly effective also for subsampling.[8] The example shown in Figure G.7 uses a dithered photo rather than a node set designed for PDE solution, since it features extremely strong density variations and thus will better convey visually any subsampling flaws. The gray levels of an original photo were here first converted to node density information which was then in turn converted to 36,303 dots using the algorithm in van der Sande and Fornberg (2021). This dense set of dots was then subsampled to 10,553 and again to 3,404 dots, using the algorithm in Lawrence et al. (2024).

The reverse process of upsampling (inserting nodes in a coarser representation) is typically less practical since each step tends to be more costly and also requires additional information not contained in the previous node set.

[8] Note also Yuksel (2015) for sample elimination in a computer graphics context.

References

Abdeljawad, T. 2015. On conformable fractional calculus. *J. Comp. Appl. Math.*, **279**, 57–66. (Cited on p. 141.)

Abreu, R., Stich, D., and Morales, J. 2015. The complex-step-finite-difference method. *Geophys. J. Int.*, **202**(1), 72–93. (Cited on p. 115.)

Ahlfors, L. V. 1966. *Complex Analysis*. McGraw-Hill. (Cited on pp. 112 and 123.)

Angel, J. B., Banks, J. W., Carson, A. M., and Henshaw, W. D. 2023. Efficient upwind finite-difference schemes for wave equations on overset grids. *SIAM J. Sci. Comput.*, **45**(5), A2703–A2724. (Cited on p. 84.)

Apostol, T. M. 1999. An elementary view of Euler's summation formula. *Am. Math. Mon.*, **106**(5), 409–418. (Cited on p. 126.)

Appelö, D., and Kreiss, G. 2006. A new absorbing layer for elastic waves. *J. Comput. Phys.*, **215**, 642–660. (Cited on p. 84.)

Appelö, D., Hagstrom, T., and Kreiss, G. 2006. Perfectly matched layers for hyperbolic systems: General formulation, well-posedness, and stability. *SIAM J. Appl. Math.*, **67**(1), 1–23. (Cited on p. 84.)

Arakawa, T., Ibukiyama, T., and Kaneko, M. 2014. *Bernoulli Numbers and Zeta Functions*. Springer Monographs in Mathematics. Springer. (Cited on p. 126.)

Ascher, U. M., and Petzold, L. R. 1988. *Computer Methods for Ordinary Differential Equations and Differential-Algebraic Equations*. SIAM. (Cited on p. 26.)

Ascher, U. M., Mattheij, R. M. M., and Russell, R. D. 1995. *Numerical Solution of Boundary Value Problems for Ordinary Differential Equations*. SIAM. (Cited on p. 26.)

Ascher, U. M., Ruuth, S. J., and Spiteri, R. J. 1997. Implicit-explicit Runge-Kutta methods for time-dependent partial differential equations. *Appl. Numer. Math.*, **25**(2–3), 151–167. (Cited on p. 67.)

Atkinson, K. E. 1988. *An Introduction to Numerical Analysis*. 2nd ed. John Wiley & Sons. (Cited on p. 126.)

Austin, A. P., Kravanja, P., and Trefethen, L. N. 2014. Numerical algorithms based on analytic function values at roots of unity. *SIAM J. Numer. Anal.*, **52**(4), 1795–1821. (Cited on p. 115.)

Balam, R. I., and Zapata, M. U. 2022. A fourth-order compact implicit immersed interface method for 2D Poisson interface problems. *Comput. Math. Appl.*, **119**, 257–277. (Cited on p. 83.)

Barton, D., Willers, I. M., and Zahar, R. V. M. 1971. An implementation of the Taylor series method for ordinary differential equations. *Comput. J.*, **14**(3), 243–248. (Cited on p. 31.)

Bayliss, A., and Turkel, E. 1980. Radiation boundary conditions for wave-like equations. *Commun. Pure Appl. Math.*, **33**(6), 707–725. (Cited on p. 84.)

Bayliss, A., Goldstein, C. I., and Turkel, E. 1983. An iterative method for the Helmholtz equation. *J. Comput. Phys.*, **49**, 443–457. (Cited on p. 82.)

Bayona, V. 2019a. Comparison of moving least squares and RBF+poly for interpolation and derivative approximation. *J. Sci. Comput.*, **81**, 486–512. (Cited on p. 86.)

Bayona, V. 2019b. An insight into RBF-FD approximations augmented with polynomials. *Comput. Math. Appl.*, **77**, 2337–2353. (Cited on pp. 99, 100, 110, and 200.)

Bayona, V., Flyer, N., and Fornberg, B. 2019. On the role of polynomials in RBF-FD approximations: III. Behavior near domain boundaries. *J. Comput. Phys.*, **380**, 378–399. (Cited on pp. 99, 100, 101, 110, 200, and 202.)

Bayona, V., Flyer, N., Fornberg, B., and Barnett, G. A. 2017. On the role of polynomials in RBF-FD approximations: II. Numerical solution of elliptic PDEs. *J. Comput. Phys.*, **332**, 257–273. (Cited on pp. 111 and 201.)

Bender, C. M., and Orszag, S. A. 1978. *Advanced Methods for Scientists and Engineers*. McGraw-Hill, 1978 (reprinted by Springer 1999). (Cited on pp. 6 and 190.)

Benjamin, T. B., and Feir, J. E. 1967. The disintegration of wave trains on deep water. Part 1. Theory. *J. Fluid Mech.*, **27**(3), 417–430. (Cited on p. 66.)

Berenger, J.-P. 1994. A perfectly matched layer for the absorption of electromagnetic waves. *J. Comput. Phys.*, **114**(2), 185–200. (Cited on p. 84.)

Berenger, J.-P. 2022. *Perfectly Matched Layer (PML) for Computational Electromagnetics*. Reprint of original edition, Morgan & Claypool 2007 ed. Synthesis Lectures on Computational Electromagnetics, no. 8. Springer. (Cited on p. 84.)

Berger, M. J., and Colella, P. 1989. Local adaptive mesh refinement for shock hydrodynamics. *J. Comput. Phys.*, **82**, 64–84. (Cited on p. 83.)

Berger, M. J., and Oliger, J. 1984. Adaptive mesh refinement for hyperbolic partial differential equations. *J. Comput. Phys.*, **53**(3), 484–512. (Cited on p. 83.)

Berger, M. J., George, D. L., LeVeque, R. J., and Mandli, K. L. 2011. The GeoClaw software for depth-averaged flows with adaptive refinement. *Adv. Water Resour.*, **34**(9), 1195–1206. (Cited on p. 83.)

Berrut, J.-P., and Trefethen, L. N. 2004. Barycentric Lagrange interpolation. *SIAM Rev.*, **46**(3), 501–517. (Cited on p. 156.)

Björck, Å., and Pereyra, V. 1970. Solution of Vandermonde systems of equations. *Math. Comput.*, **24**(112), 893–903. (Cited on p. 147.)

Bochner, S. 1933. Monotone Functionen, Stieltjes Integrale und Harmonische Analyse. *Math. Ann.*, **108**, 378–410. (Cited on p. 93.)

Bollig, E., Flyer, N., and Erlebacher, G. 2012. Solution to PDEs using radial basis function finite differences (RBF-FD) on multiple GPUs. *J. Comput. Phys.*, **231**, 7133–7151. (Cited on p. 105.)

Bornemann, F. 2011. Accuracy and stability of computing high-order derivatives of analytic functions by Cauchy integrals. *Found. Comput. Math.*, **11**, 1–63. (Cited on p. 116.)

Borwein, J. M., Calkin, N. J., and Manna, D. 2009. Euler–Boole summation revisited. *Am. Math. Mon.*, **116**(5), 387–412. (Cited on p. 137.)

Boyd, J. P. 2000. *Chebyshev and Fourier Spectral Methods*. Dover. (Cited on pp. 17 and 68.)

Boyd, J. P., and Flyer, N. 1999. Compatibility conditions for time-dependent partial differential equations and the rate of convergence of Chebyshev and Fourier spectral methods. *Comput. Meth. Appl. Mech. Eng.*, **175**, 281–309. (Cited on p. 65.)

Bradie, B. 2006. *A Friendly Introduction to Numerical Analysis*. Pearson Prentice Hall. (Cited on p. 42.)

Brandt, A. 1977. Multi-level adaptive solutions to boundary-value problems. *Math. Comput.*, **31**, 333–390. (Cited on p. 82.)

Brezinski, C. 1991. *History of Continued Fractions and Padé Approximants*. Springer Series in Computational Mathematics, vol. 12. Springer. (Cited on p. 190.)

Brezinski, C. 2009. *The Birth of Numerical Analysis*. Chapter 1: Some pioneers of extrapolation methods. World Scientific. (Cited on p. 185.)

Brezinski, C., and Redivo-Zaglia, M. 1991. *Extrapolation Methods: Theory and Practice*. North Holland. (Cited on pp. 138, 185, and 190.)

Brezinski, C., and Redivo-Zaglia, M. 2020. *Extrapolation and Rational Approximation: The Works of the Main Contributors*. Springer Nature Switzerland. (Cited on p. 190.)

Briggs, W. L., Henson, V. E., and McCormick, S. F. 2000. *A Multigrid Tutorial*. 2nd ed. SIAM. (Cited on p. 82.)

Brown, J. W., and Churchill, R. V. 2013. *Complex Variables and Applications*. 9th ed. McGraw Hill. (Cited on p. 112.)

Bungartz, H.-J., and Griebel, M. 2004. Sparse grids. *Acta Numerica*, **13**, 147–269. (Cited on p. 205.)

Burden, R. L., and Faires, J. D. 2015. *Numerical Analysis*. 10th ed. Brooks/Cole – Cengage Learning. (Cited on p. 42.)

Burkardt, J., Wu, Y., and Zhang, Y. 2021. A unified meshfree pseudospectral method for solving both classical and fractional PDEs. *SIAM J. Sci. Comput.*, **42**(2), A1389–A1411. (Cited on p. 154.)

Butcher, J. C. 1996. A history of Runge-Kutta methods. *Appl. Numer. Math.*, **20**, 247–260. (Cited on p. 29.)

Butcher, J. C. 2016. *Numerical Methods for Ordinary Differential Equations*. 3rd ed. Wiley. (Cited on pp. 26 and 27.)

Caflisch, R. E. 1998. Monte Carlo and quasi-Monte Carlo methods. *Acta Numer.*, **7**, 1–49. (Cited on p. 206.)

Calhoun, D. 2002. A Cartesian grid method for solving the two-dimensional streamfunction-vorticity equation in irregular regions. *J. Comput. Phys.*, **176**, 231–275. (Cited on p. 83.)

Caputo, M. 1967. Linear model of dissipation whose Q is almost frequency independent – II. *Geophys. J. R. Astron. Soc.*, **13**, 529–539. (Cited on p. 142.)

Cardone, A., Jackiewicz, Z., Sandu, A., and Zhang, H. 2014. Extrapolation-based implicit-explicit general linear methods. *Numer. Algor.*, **65**, 377–399. (Cited on p. 67.)

Carpenter, M. H., Gottlieb, D., Abarbanel, S., and Don, W.-S. 1995. The theoretical accuracy of Runge-Kutta time discretization for the initial boundary value problem: A study of the boundary error. *SIAM J. Sci. Comput.*, **16**(6), 1241–1252. (Cited on p. 50.)

Chang, Y. F., and Corliss, G. 1994. ATOMFT: Solving ODEs and DAEs using Taylor series. *Comp. Math. Applic.*, **28**(10–12), 209–233. (Cited on p. 31.)

Cheney, W., and Kincaid, D. 2012. *Numerical Mathematics and Computing*. 7th ed. Cengage Learning. (Cited on p. 42.)

Chesshire, G., and Henshaw, W. D. 1990. Composite overlapping meshed for the solution of partial differential equations. *J. Comput. Phys.*, **90**(1), 1–64. (Cited on p. 84.)

Chu, T., and Schmidt, O. T. 2023. RBF-FD discretization of the Navier-Stokes equations using staggered nodes. *J. Comput. Phys.*, **474**, Article Nr: 111756. (Cited on p. 109.)

Ciesielski, M. 2024. Numerical algorithms for approximation of fractional integrals and derivatives based on quintic spline interpolation. *Symmetry*, **16**, Article Nr: 252. (Cited on p. 148.)

Collatz, L. 1960. *The Numerical Treatment of Differential Equations*. Springer-Verlag. (Cited on p. 72.)

Cooley, J. W., and Tukey, J. W. 1965. An algorithm for the machine calculation of complex Fourier series. *Math. Comput.*, **19**, 297–301. (Cited on p. 177.)

Courant, R., Friedrichs, K., and Lewy, H. 1928. On the partial differential equations of mathematical physics. *Math. Ann.*, **100**, 32–74 (in German; reproduced in English in IBM Journal of Research and Development 11 (2), 1967, 215–234). (Cited on p. 57.)

Dahlquist, G. 1956. Convergence and stability in the numerical integration of ordinary differential equations. *Math. Scand.*, **4**, 33–53. (Cited on p. 41.)

Dahlquist, G. 1963. A special stability problem for linear multistep methods. *BIT*, **3**, 27–43. (Cited on p. 41.)

Daoud, M., and Laamri, E. H. 2022. Fractional Laplacian: A short survey. *Dis. Cont. Dynam. Syst. S*, **15**(1), 95–116. (Cited on p. 150.)

de Boor, C. 1962. Bicubic spline interpolation. *J. Mathematics and Physics*, **41**(1–4), 212–218. (Cited on p. 165.)

de Boor, C. 2001. *A Practical Guide to Splines*. Revised ed. Springer. (Cited on pp. 159 and 164.)

D'Elia, M., Du, Q., Glusa, C., Gunzburger, M., Tian, X., and Zhou, Z. 2020. Numerical methods for nonlocal and fractional models. *Acta Numerica*, **29**, 1–124. (Cited on p. 150.)

Demmel, J., and Koev, P. 2005. The accurate and efficient solution of a totally positive generalized Vandermonde linear system. *SIAM J. Matrix Anal. Appl.*, **27**(1), 142–152. (Cited on p. 147.)

Deshpande, V. M., Bhattacharya, R., and Donzis, D. A. 2021. A unified framework to generate optimized compact finite difference schemes. *J. Comput. Phys.*, **432**, Article Nr: 110157. (Cited on p. 195.)

Deuflhard, P., and Bornemann, F. 2002. *Scientific Computing with Ordinary Differential Equations*. Texts in Applied Mathematics, vol. 42. Springer. (Cited on p. 26.)

Dierckx, P. 1993. *Fitting with Splines*. Oxford University Press. (Cited on p. 159.)

Diethelm, K. 2010. *The Analysis of Fractional Differential Equations*. Springer. (Cited on p. 141.)

Dragomir, S. S., Agarwal, R. P., and Cerone, P. 2000. On Simpson's inequality and applications. *J. Ineq. App.*, **5**(6), 533–579. (Cited on p. 140.)

Driscoll, T. A., and Fornberg, B. 1999. Block pseudospectral methods for Maxwell's equations II: Two-dimensional, discontinuous-coefficient case. *SIAM J. Sci. Comput.*, **21**(3), 1146–1167. (Cited on p. 84.)

Driscoll, T. A., and Fornberg, B. 2001. A Padé-based algorithm for overcoming the Gibbs phenomenon. *Num. Algor.*, **26**, 77–92. (Cited on p. 23.)

Driscoll, T. A., and Fornberg, B. 2002. Interpolation in the limit of increasingly flat radial basis functions. *Comput. Math. Applic.*, **43**, 413–422. (Cited on p. 94.)

Dutt, A., and Rokhlin, V. 1993. Fast Fourier transforms for nonequispaced data. *SIAM J. Sci. Comput.*, **14**(6), 1368–1393. (Cited on p. 179.)

Eckhoff, K. S. 1998. On a high order numerical method for functions with singularities. *Math. Comput.*, **67**, 1063–1087. (Cited on p. 23.)

Ellison, A. C., and Fornberg, B. 2021. A parallel-in-time approach for wave-type PDEs. *Numerische Mathematik*, **148**, 79–98. (Cited on pp. 200, 202, and 203.)

Engquist, B., and Majda, A. 1979. Radiation boundary conditions for acoustic and elastic wave calculations. *Comm. Pure Appl. Math.*, **32**(3), 312–358. (Cited on p. 84.)

Ernst, O. G., and Gander, M. J. 2012. Why it is difficult to solve Helmholtz problems with classical iterative methods. Pages 325–363 of: Graham, I., Hou, T., Lakkis, O., and Scheichl, R. (eds.), *Numerical Analysis of Multiscale Problems*. Lecture Notes in Computational Science and Engineering, vol. 83. Springer. (Cited on p. 82.)

Falgout, R. D. 2006. An introduction to algebraic multigrid. *Comp. Sci. Engin.*, **8**(6), 24–33. (Cited on p. 83.)

Fasondini, M., Fornberg, B., and Weideman, J. A. C. 2017. Methods for the computation of the multivalued Painlevé transcendents on their Riemann surfaces. *J. Comput. Phys.*, **344**, 36–50. (Cited on p. 30.)

Fasshauer, G. E. 2007. *Meshfree Approximation Methods with MATLAB*. Interdisciplinary Mathematical Sciences 6. World Scientific Publishers. (Cited on pp. 89, 90, and 93.)

Fasshauer, G. E., and McCourt, M. J. 2012. Stable evaluation of Gaussian radial basis function interpolants. *SIAM J. Sci. Comput.*, **34**(2), A737–A762. (Cited on p. 95.)

Fathi, A., Poursartip, B., and Kallivokas, L. F. 2015. Time-domain hybrid formulations for wave simulations in three-dimensional PML-truncated heterogeneous media. *Num. Meth. in Eng.*, **101**(3), 165–198. (Cited on p. 84.)

Fedoseyev, A. I., Friedman, M. J., and Kansa, E. J. 2002. Improved multiquadric method for elliptic partial differential equations via PDE collocation on the boundary. *Comput. Math. Applic.*, **43**(3), 439–455. (Cited on p. 111.)

Ferraro, G. 2008. *The Rise and Development of the Theory of Series up to the Early 1820s*. Sources and Studies in the History of Mathematics and Physical Sciences. Springer. (Cited on p. 126.)

Floater, M. S., and Hormann, K. 2007. Barycentric rational interpolation with no poles and high rates of approximation. *Numerische Mathematik*, **107**(2), 315–331. (Cited on p. 156.)

Flyer, N., and Fornberg, B. 2003. On the nature of initial-boundary value solutions for dispersive equations. *SIAM J. Appl. Math.*, **64**(2), 546–564. (Cited on p. 65.)

Flyer, N., Lehto, E., Blaise, S., Wright, G. B., and St-Cyr, A. 2012. A guide to RBF-generated finite differences for nonlinear transport: Shallow water simulations on a sphere. *J. Comput. Phys.*, **231**, 4078–4095. (Cited on p. 106.)

Flyer, N., Fornberg, B., Bayona, V., and Barnett, G. A. 2016. On the role of polynomials in RBF-FD approximations: I. Interpolation and accuracy. *J. Comput. Phys.*, **321**, 21–38. (Cited on p. 104.)

Forest, E., and Ruth, R. D. 1990. Fourth order symplectic integration. *Phys. D*, **43**, 105–117. (Cited on p. 68.)

Fornberg, B. 1973. On the instability of leap-frog and Crank-Nicolson approximations of a nonlinear partial differential equation. *Math. Comput.*, **27**, 45–57. (Cited on p. 21.)

Fornberg, B. 1981. Numerical differentiation of analytic functions. *ACM Trans. Math. Software*, **7**(4), 512–526. (Cited on pp. 115 and 116.)

Fornberg, B. 1987. The pseudospectral method: Comparisons with finite differences for the elastic wave equation. *Geophysics*, **52**(4), 483–501. (Cited on pp. 19 and 20.)

Fornberg, B. 1988a. Generation of finite difference formulas on arbitrarily spaced grids. *Math. Comput.*, **51**(184), 699–706. (Cited on p. 5.)

Fornberg, B. 1988b. Steady viscous flow past a sphere at high Reynolds numbers. *J. Fluid Mech.*, **190**, 471–489. (Cited on p. 84.)

Fornberg, B. 1990. High-order finite differences and the pseudospectral method on staggered grids. *SIAM J. Num. Anal.*, **27**(4), 904–918. (Cited on p. 19.)

Fornberg, B. 1996. *A Practical Guide to Pseudospectral Methods*. Cambridge Monographs on Applied and Computational Mathematics, vol. 1. Cambridge University Press. (Cited on pp. 13, 17, 23, 40, 50, 57, 63, and 158.)

Fornberg, B. 1998. Calculation of weights in finite difference formulas. *SIAM Rev.*, **40**(3), 685–691. (Cited on p. 5.)

Fornberg, B. 2010. A finite difference method for free boundary problems. *J. Comp. Appl. Math.*, **233**, 2831–2840. (Cited on p. 83.)

Fornberg, B. 2020. Euler-Maclaurin expansions without analytic derivatives. *Proc. Royal Soc. London, Series A*, **476**, Article Nr: 20200441. (Cited on pp. 126, 131, and 132.)

Fornberg, B. 2021a. An algorithm for calculating Hermite-based finite difference weights. *IMA J. Numer. Anal.*, **41**, 801–813. (Cited on pp. 14 and 156.)

Fornberg, B. 2021b. Contour integrals of analytic functions given on a grid in the complex plane. *IMA J. Numer. Anal.*, **41**, 814–825. (Cited on pp. 133, 135, and 136.)

Fornberg, B. 2021c. *Encyclopedia of Mathematical Geosciences*. Encyclopedia of Earth Sciences. ed. B. S. Daya Sagar et. al. Springer. Chap. Splines. (Cited on pp. 159, 161, 163, 164, and 165.)

Fornberg, B. 2021d. Generalizing the trapezoidal rule in the complex plane. *Num. Algor.*, **87**, 187–202. (Cited on p. 136.)

Fornberg, B. 2021e. Improving the accuracy of the trapezoidal rule. *SIAM Rev.*, **63**(1), 167–180. (Cited on p. 197.)

Fornberg, B. 2022. Finite difference formulas in the complex plane. *Num. Algor.*, **90**, 1305–1326. (Cited on pp. 78 and 118.)

Fornberg, B. 2023. Infinite order accuracy limit of finite difference formulas in the complex plane. *IMA J. Num. Anal.*, **43**, 3055–3072. (Cited on pp. 119, 120, and 134.)

Fornberg, B., and Driscoll, T. A. 1999. A fast spectral algorithm for nonlinear wave equations with linear dispersion. *J. Comput. Phys.*, **155**, 456–467. (Cited on p. 68.)

Fornberg, B., and Elcrat, A. 2014. Some observations regarding steady laminar flows past bluff bodies. *Phil. Trans. Royal Soc. A.*, **372**, Article Nr: 20130353. (Cited on p. 70.)

Fornberg, B., and Flyer, N. 2015a. Fast generation of 2-D node distributions for mesh-free PDE discretizations. *Comp. Math. Applic.*, **69**, 531–544. (Cited on pp. 207 and 208.)

Fornberg, B., and Flyer, N. 2015b. *A Primer on Radial Basis Functions with Applications to the Geosciences*. CBMS-NSF, vol. 87. SIAM. (Cited on pp. 23, 39, 40, 90, 93, 94, 106, and 206.)

Fornberg, B., and Flyer, N. 2015c. Solving PDEs with radial basis functions. *Acta Numerica*, **24**, 215–258. (Cited on pp. 90 and 92.)

Fornberg, B., and Lawrence, A. P. 2023. Enhanced trapezoidal rule for discontinuous functions. *J. Comput. Phys.*, **491**, Article Nr: 112386. (Cited on pp. 197 and 198.)

Fornberg, B., and Lehto, E. 2011. Stabilization of RBF-generated finite difference methods for convective PDEs. *J. Comput. Phys.*, **230**, 2270–2285. (Cited on pp. 106, 107, 108, and 109.)

Fornberg, B., and Piret, C. 2007. A stable algorithm for flat radial basis functions on a sphere. *SIAM J. Sci. Comput.*, **30**(1), 60–80. (Cited on p. 95.)

Fornberg, B., and Piret, C. 2020. *Complex Variables and Analytic Functions: An Illustrated Introduction*. SIAM. (Cited on pp. 6, 30, 35, 112, 116, 123, 139, and 152.)

Fornberg, B., and Piret, C. 2023. Computation of fractional derivatives of analytic functions. *J. Sci. Comput.*, **96**, Article Nr: 79. (Cited on pp. 142, 144, 149, and 150.)

Fornberg, B., and Weideman, J. A. C. 2011. A numerical method for the Painlevé equations. *J. Comput. Phys.*, **230**, 5957–5973. (Cited on pp. 30 and 133.)

Fornberg, B., and Whitham, G. B. 1978. A numerical and theoretical study of certain nonlinear wave phenomena. *Phil. Trans. Royal Soc. Ser. A*, **289**, 373–404. (Cited on pp. 67 and 68.)

Fornberg, B., and Wright, G. 2004. Stable computation of multiquadric interpolants for all values of the shape parameter. *Comput. Math. Applic.*, **48**, 853–867. (Cited on p. 95.)

Fornberg, B., Zuev, J., and Lee, J. 2007. Stability and accuracy of time-extrapolated ADI-FDTD methods for solving wave equations. *J. Comput. Appl. Math.*, **200**, 178–192. (Cited on p. 201.)

Fornberg, B., Flyer, N., and Russell, J. M. 2010. Comparisons between pseudospectral and radial basis function derivative approximations. *IMA J. Numer. Anal.*, **30**, 149–172. (Cited on pp. 24 and 25.)

Fornberg, B., Larsson, E., and Flyer, N. 2011. Stable computations with Gaussian radial basis functions. *SIAM J. Sci. Comput.*, **33**(2), 869–892. (Cited on p. 95.)

Fornberg, B., Lehto, E., and Powell, C. 2013. Stable calculation of Gaussian-based RBF-FD stencils. *Comp. Math. Applic.*, **65**, 627–637. (Cited on p. 95.)

Fornberg, B., Piret, C., and Higgins, A. 2025. *The Fundamentals of Fractional Calculus*. AAP / CRC Press. Chap. 6. Numerical computation of fractional derivatives of Caputo type. Eds: D. K. Singh and M. Yavuz. (Cited on p. 142.)

Fox, L. 1947. Some improvements in the use of relaxation methods for the solution of ordinary and partial differential equations. *Proc. R. Soc. London, Ser. A*, **190**, 31–59. (Cited on pp. 45 and 72.)

Gander, M. J. 2015. 50 years of time parallel time integration. Pages 69–113 of: Carraro, T., Geiger, M., Körkel, S., and Rannacher, R. (eds.), *in Multiple Shooting and Time Domain Decomposition Methods*. Springer. (Cited on pp. 29 and 200.)

Gardner, C. S., Green, J. M., Kruskal, M. D., and Miura, R. M. 1967. Method for solving the Korteweg-deVries equation. *Phys. Rev. Lett.*, **19**, 1095–1097. 1967. (Cited on p. 66.)

Garofalo, N. 2019. Fractional thoughts. Pages 1–135 of: Petroysan, A., Danielli, D. and Pop, C. A. (eds.), *New Developments in the Analysis of Nonlocal Operators*. Contemporary Mathematics, vol. 723. American Mathematical Society. (Cited on p. 150.)

Gavete, L., Gavete, M. L., and Benito, J. J. 2003. Improvements of generalized finite difference method and comparison with other meshless method. *Appl. Math. Modelling*, **27**, 831–847. (Cited on p. 85.)

Gerstner, T., and Griebel, M. 1998. Numerical integration using sparse grids. *Numerical Algorithms*, **18**, 209–232. (Cited on p. 205.)

Ghrist, M., Fornberg, B., and Driscoll, T. A. 2000. Staggered time integrators for wave equations. *SIAM J. Num. Anal.*, **38**(3), 718–741. (Cited on p. 58.)

Ghrist, M. L., Fornberg, B., and Reeger, J. A. 2015. Stability ordinates of Adams predictor-corrector methods. *BIT*, **55**, 733–750. (Cited on p. 40.)

Gibou, F., Fedkiw, R., and Osher, S. 2018. A review of level-set methods and some recent applications. *J. Comput. Phys.*, **353**, 82–109. (Cited on p. 84.)

Giles, M. B. 2015. Multilevel Monte Carlo methods. *Acta Numerica*, **24**, 259–328. (Cited on p. 206.)

Giraldo, F. X., Kelly, J. F., and Constantinescu, E. M. 2013. Implicit-explicit formulations of a three-dimensional nonhydrostatic unified model of the atmosphere (NUMA). *SIAM J. Sci. Comput.*, **35**(5), B1162–B1194. (Cited on p. 67.)

Givoli, D. 2023. Dahlquist's barriers and much beyond. *J. Comput. Phys.*, **475**, Article Nr: 111836. (Cited on p. 41.)

Glassman, J. A. 1970. A generalization of the fast Fourier transform. *IEEE Trans. Comput.*, **C-19**, 105–116. (Cited on p. 179.)

Glaubitz, J., Klein, S. C., Nordström, J., and Öffner, P. 2023. Multi-dimensional summation-by-parts operators for general functioon spaces: Theory and construction. *J. Comput. Phys.*, **491**, Article Nr: 112370. (Cited on p. 57.)

Glusa, C., Antil, H., D'elia, M., Waanders, B. V., and Weiss, C. J. 2021. A fast solver for the fractional order Helmholtz equation. *SIAM J. Sci. Comput.*, **43**(2), A1362–A1388. (Cited on p. 151.)

Golub, G. H., and Loan, C. F. Van. 2013. *Matrix Computations*. 4th ed. Johns Hopkins University Press. (Cited on p. 82.)

Gonnet, P., Güttel, S., and Trefethen, L. N. 2013. Robust Padé approximation via SVD. *SIAM Review*, **55**(1), 101–117. (Cited on p. 190.)

Gorenflo, R., and Mainardi, F. 2008. Fractional calculus: Integral and differential equations of fractional order. ArXiv:0805.3823v1. (Cited on p. 142.)

Gottlieb, D., and Shu, C.-W. 1997. On the Gibbs phenomenon and its resolution. *SIAM Review*, **39**, 644–668. (Cited on p. 23.)

Gottlieb, S., Ketcheson, D., and Shu, C.-W. 2011. *Strong Stability Preserving Runge-Kutta and Multistep Time Discretizations*. World Scientific. (Cited on p. 71.)

Gragg, W. B. 1965. On extrapolation algorithms for ordinary initial value problems. *SIAM J. Num. Anal.*, **2**, 384–404. (Cited on p. 36.)

Greengard, L., and Lee, J.-Y. 2004. Accelerating the nonuniform fast Fourier transform. *SIAM Rev.*, **46**(3), 443–454. (Cited on p. 179.)

Gregory, J. 1670. Letter to J. Collins, 23 November 1670. Reprinted in *Correspondence of Scientific Men*, S. Rigaud. *Oxford University Press*, pp. 203–212, (1841). (Cited on p. 128.)

Griewank, A., and Walther, A. 2008. *Evaluating Derivatives: Principles and Techniques of Algorithmic Differentiation*. 2nd ed. SIAM. (Cited on p. 31.)

Gunderman, D., Flyer, N., and Fornberg, B. 2020. Transport schemes in spherical geometries using spline-based RBF-FD with polynomials. *J. Comput. Phys.*, **408**, Article Nr: 109256. (Cited on p. 110.)

Gupta, M. M. 1991. High accuracy solutions of incompressible Navier–Stokes equations. *J. Comput. Phys.*, **93**, 343–359. (Cited on p. 73.)

Gupta, M. M., Manohar, R. P., and Stephenson, J. W. 1984. A single cell high-order scheme for the convection diffusion equation with variable coefficients. *Int. J. Num. Meth. Fluids*, **4**(7), 641–651. (Cited on p. 71.)

Gustafsson, B. 2008. *High Order Difference Methods for Time Dependent PDE*. Springer Series in Computational Mathematics, vol. 38. Springer. (Cited on pp. 26, 51, 57, and 71.)

Gustafsson, B., Kreiss, H.-O., and Oliger, J. 2013. *Time-Dependent Problems and Difference Methods*. 2nd ed. Pure and Applied Mathematics. John Wiley & Sons. (Cited on pp. 21 and 57.)

Guzmán, P. M., Langton, G., Bittencurt, L. M. Lugo Motta, Medina, J., and Valdes, J. E. Nápoles. 2018. A new definition of a fractional derivative of local type. *J. Math. Anal.*, **9**(2), 88–98. (Cited on p. 141.)

Hairer, E., and Wanner, G. 2002. *Solving Ordinary Differential Equations II: Stiff and Differential-Algebraic Problems*. 2nd ed. Springer. (Cited on p. 26.)

Hairer, E., Nørsett, S. P., and Wanner, G. 1992. *Solving Ordinary Differential Equations I: Nonstiff Problems*. 2nd revised ed. Springer. (Cited on pp. 26, 31, and 36.)

Halko, N., Martinsson, P. G., and Tropp, J. A. 2011. Finding structure with randomness: Probabilistic algorithms for constructing approximate matrix decompositions. *SIAM Rev.*, **53**(3), 217–288. (Cited on p. 81.)

Hardy, R. L. 1971. Multiquadric equations of topography and other irregular surfaces. *J. Geophys. Res.*, **76**, 1905–1915. (Cited on p. 90.)

Harlow, F. H., and Welch, J. E. 1965. Numerical calculation of time-dependent viscous incompressible flow of fluid with free surface. *Phys. Flu.*, **8**(12), 2182–2189. (Cited on pp. 58 and 108.)

Henshaw, W. D., and Schwendeman, D. W. 2006. Moving overlapping grids with adaptive mesh refinement for high-speed reactive and non-reactive flow. *J. Comput. Phys.*, **216**, 744–779. (Cited on p. 84.)

Hestens, M. R., and Stiefel, E. 1952. Methods of conjugate gradients for solving linear systems. *J. Res. National Bureau of Standards*, **49**(6), 409–436. (Cited on p. 82.)

Hesthaven, J. S. 2018. *Numerical Methods for Conservation Laws*. SIAM. (Cited on p. 71.)

Higgins, A. 2022. Numerical computations of fractional derivatives of analytic functions. *SIAM Undergrad. Res. Online*, **15**, 511–525. (Cited on p. 150.)

Higham, N. J. 2004. The numerical stability of barycentric Lagrange interpolation. *IMA J. Numer. Anal.*, **24**(4), 547–556. (Cited on p. 156.)

Hiptmair, R., Moiola, A., and Perugia, I. 2016. A survey of Trefftz methods for the Helmholtz equation. Pages 237–279 of: Barrenechea, G. R., Brezzi, F., Cangiani, A., and Georgoulis, E. H. (eds.), *Building Bridges: Connections and Challenges in Modern Approaches to Numerical Patrial Differential Equations*. Lecture Notes in Computational Science and Engineering, vol. 114. Springer. (Cited on p. 76.)

Hochbruck, M., and Ostermann, A. 2010. Exponential integrators. *Acta Numerica*, **19**, 209–286. (Cited on p. 67.)

Hofreither, C. 2020. A unified view of some numerical methods for fractional diffusion. *Comp. Math. Applic.*, **80**, 332–350. (Cited on p. 151.)

Hofreither, C. 2021. An algorithm for best rational approximation based on barycentric rational interpolation. *Num. Algor.*, **88**, 365–388. (Cited on p. 156.)

Holberg, O. 1987. Computational aspects of the choice of operator and sampling interval for numerical differentiation in large-scale simulation of wave phenomena. *Geo. Prosp.*, **35**(6), 629–655. (Cited on p. 195.)

Hung, Y.-C. 2023. A review of Monte Carlo and quasi-Monte Carlo sampling techniques. *Wiley Interdisciplinary Reviews (WIREs)*. (Cited on p. 206.)

Iserles, A. 2012. *A First Course in the Numerical Analysis of Differential Equations*. 2nd ed. Cambridge Texts in Applied Mathematics, vol. 44. Cambridge University Press. (Cited on pp. 26 and 49.)

Iserles, A., and Nørsett, S. P. 1990. On the theory of parallel Runge-Kutta methods. *IMA J. Numer. Anal.*, **10**(4), 463–488. (Cited on p. 29.)

Ishteva, M. K. 2005. *Properties and applications of the Caputo fractional operator*. M.Phil. thesis, University of Karlsruhe. (Cited on p. 142.)

Iske, A. 2003. On the approximation order and numerical stability of local Lagrange interpolation by polyharmonic splines. Haussman, W., Jetter, K., Reimer, M., and Stöckler, J. (eds.), *in Modern Developments in Multivariate Approximation*. International Series of Numerical Mathematics, vol. 145. Birkhäuser Verlag. (Cited on p. 104.)

Jensen, P. S. 1972. Finite difference techniques for variable grids. *Comp. Struct.*, **2**, 17–29. (Cited on pp. 85 and 86.)

Johnson, S. G. 2021. Notes on perfectly matched layers (PMLs). ArXiv: 2108.05348v1. (Cited on p. 84.)

Kansa, E. J. 1990a. Multiquadrics – a scattered data approximation scheme with applications to computational fluid dynamics I: Surface approximations and partial derivative estimates. *Comp. Math. Applic.*, **19**, 127–145. (Cited on p. 92.)

Kansa, E. J. 1990b. Multiquadrics – a scattered data approximation scheme with applications to computational fluid dynamics II: Solutions to parabolic, hyperbolic and elliptic partial differential equations. *Compt. Math. Applic.*, **19**, 147–161. (Cited on p. 92.)

Kassam, A. K., and Trefethen, L. N. 2005. Fourth-order time-stepping for stiff PDEs. *SIAM J. Sci. Comput.*, **26**(4), 1214–1233. (Cited on p. 67.)

Ketcheson, D. I., and Ahmadia, A. J. 2012. Optimal stability polynomials for numerical integration of initial value problems. *Commun. Appl. Math. Comput. Sci.*, **7**, 247–271. (Cited on p. 202.)

Ketcheson, D. I., and BinWaheed, U. 2014. A comparison of high-order explicit Runge-Kutta, extrapolation, and deferred correction methods in serial and parallel. *Commun. Appl. Math. Comput. Sci.*, **9**, 175–200. (Cited on p. 29.)

Khalil, R., Horani, M. Al, Yousef, A., and Sababheh, M. 2014. A new definition of fractional derivative. *J. Comp. Appl. Math.*, **264**, 65–70. (Cited on p. 141.)

Kilbas, A. A., Srivastava, H. M., and Trujillo, J. I. 2006. *Theory and Applications of Fractional Differential Equations*. Elsevier. (Cited on p. 142.)

Kinnmark, I. P. E., and Gray, W. G. 1984. One step integration methods with maximum stability regions. *Math. Comput. Simul.*, **26**, 87–92. (Cited on p. 202.)

Kreiss, H.-O., and Oliger, J. 1972. Comparison of accurate methods for the integration of hyperbolic equations. *Tellus*, **24**, 199–215. (Cited on pp. 19 and 20.)

Kreiss, H.-O., and Ortiz, O. E. 2014. *Introduction to Numerical Methods for Time Dependent Differential Equations*. Wiley. (Cited on p. 49.)

Krogh, F. T. 1966. Predictor-corrector methods of high order with improved stability characteristics. *J. Assoc. Comp. Mach.*, **13**(3), 374–385. (Cited on p. 40.)

Krusemeyer, M. 1994. *Differential Equations*. Macmillan College Publishing. (Cited on p. 68.)

Kumari, K., Bhattacharya, R., and Donzis, D. A. 2019. A unified approach for deriving optimal finite differences. *J. Comput Phys.*, **399**, Article Nr: 108957. (Cited on p. 195.)

Kunz, K. S., and Lubbers, R. J. 1993. *The Finite Difference Time Domain Method for Electromagnetics*. CRC Press. (Cited on p. 61.)

Lambe, L. A., Luczak, R., and Nehrbass, J. W. 2003. A new finite difference method for the Helmholtz equation using symbolic computation. *Int. J. Comp. Eng, Sci.*, **4**(1), 121–144. (Cited on p. 76.)

Lambert, J. D. 1991. *Numerical Methods for Ordinary Differential Systems: The Initial Value Problem*. 1st ed. Wiley. (Cited on pp. 26, 27, and 36.)

Lancaster, P., and Salkauskas, K. 1981. Surfaces generated by moving least squares methods. *Math. Comput.*, **37**(155), 141–158. (Cited on p. 88.)

Lanzara, F., Maz'ya, V., and Schmidt, G. 2022. Fast computation of the multidimensional fractional Laplacian. *Applic. Anal.*, **101**(11), 4025–4041. (Cited on p. 153.)

Larsson, E., and Fornberg, B. 2003. A numerical study of some radial basis function based solution methods for elliptic PDEs. *Comp. Math. Applic.*, **46**(5), 891–902. (Cited on p. 111.)

Lawrence, A. P., Nielsen, M. E., and Fornberg, B. 2024. Node subsampling for multilevel meshfree elliptic PDE solvers. *Comp. Math. Applic.*, **164**, 79–94. (Cited on pp. 111 and 210.)

Lax, P. D. 1967. Hyperbolic difference equations: A review of the Courant-Friedrichs-Lewy paper in the light of recent developments. *IBM J.*, **11**, 235–238. (Cited on p. 57.)

Lax, P. D., and Richtmyer, R. D. 1956. Survey of the stability of linear finite difference equations. *Comm. Pure Appl. Math.*, **9**(2), 267–293. (Cited on p. 51.)

Lee, J., and Fornberg, B. 2003. A split step approach for the 3-D Maxwell's equations. *J. Comput. Appl. Math.*, **158**, 485–505. (Cited on p. 68.)

Leinen, W. 2024. *Iterative solvers for RBF-FD discretized flow problems*. Ph.D. thesis, Technical University of Hamburg. (Cited on p. 111.)

Lele, S. K. 1992. Compact finite difference schemes with spectral-like resolution. *J. Comput. Phys.*, **103**, 16–42. (Cited on p. 71.)

LeVeque, R. J. 1990, 2nd ed. 2008. *Numerical Methods for Conservation Laws*. Birkhäuser. (Cited on p. 71.)

LeVeque, R. J. 2002. *Finite Volume Methods for Hyperbolic Problems*. Cambridge Texts in Applied Mathematics, vol. 31. Cambridge University Press. (Cited on pp. 58 and 71.)

LeVeque, R. J. 2007. *Finite Difference Methods for Ordinary and Partial Differential Equations*. SIAM. (Cited on pp. 26 and 49.)

Li, C., and Zeng, F. 2015. *Numerical Methods for Fractional Calculus*. CRC Press. (Cited on p. 144.)

Li, M., Tang, T., and Fornberg, B. 1995. A compact fourth-order finite difference scheme for the steady incompressible Navier-Stokes equations. *Int. J. Numer. Meth. Fluids*, **20**, 1137–1151. (Cited on p. 71.)

Li, Z., and Ito, K. 2006. *The Immersed Interface Method*. SIAM. (Cited on p. 83.)

Liouville, J. 1832. Memóire sur le calcul des différentielles à indices quelconques. *J. de l'École Polytechnique, Paris*, **13**, 71–162. (Cited on p. 141.)

Lipnikov, K., Manzini, G., and Shaskow, M. 2014. Mimetic finite difference method. *J. Comput. Phys.*, **257**, 1163–1227. (Cited on p. 194.)

Lischke, A., Pang, G., Gulian, M., Song, F., Glusa, C., Zheng, X., Mao, Z., Cai, W., Meerschaert, M. M., Ainsworth, M., and Karniadakis, G.Em. 2020. What is the fractional Laplacian? A comparative review with new results. *J. Comput. Phys.*, **404**, Article Nr: 109009. (Cited on p. 151.)

Liszka, T., and Orkisz, J. 1980. The finite difference method at arbitrary irregular grids and its application in applied mechanics. *Comp. Struct.*, **11**, 83–95. (Cited on p. 85.)

Lynch, R. E., and Rice, J. R. 1978. High accuracy finite difference approximations to solutions of elliptic partial differential equations. *Proc. Natl. Acad. Sci. USA*, **75**(6), 2541–2544. (Cited on p. 73.)

Lyness, J. N., and Moler, C. B. 1967. Numerical differentiation of analytic functions. *SIAM J. Numer. Anal.*, **4**, 202–210. (Cited on p. 115.)

Lyness, J. N., and Sande, G. 1971. Algorithm 413-ENTCAF and ENTCRE: Evaluation of normalized Taylor coefficients of an analytic function. *Comm. ACM*, **14**(10), 669–675. (Cited on p. 115.)

Manohar, R., and Stephenson, J. W. 1984. High order difference schemes for linear partial differential equations. *SIAM J. Sci. Stat. Comput.*, **5**(1), 69–77. (Cited on p. 73.)

Martensen, E. 1964. Optimale Fehlerschranken für die Qadraturformel von Gregory. *ZAMM*, **44**(4/5), 159–168. (Cited on p. 139.)

Martin, B., and Fornberg, B. 2017a. Seismic modeling with radial basis function-generated finite differences (RBF-FD) – a simplified treatment of interfaces. *J. Comput. Phys.*, **335**, 828–845. (Cited on p. 110.)

Martin, B., and Fornberg, B. 2017b. Using radial basis function-generated finite differences (RBF-FD) to solve heat transfer equilibrium problems in domains with interfaces. *Eng. Anal. with Boundary Elements*, **79**, 38–48. (Cited on p. 110.)

März, T., and Macdonald, C. B. 2012. Calculus on surfaces with general closest point functions. *SIAM J. Numer. Anal.*, **50**, 3303–3328. (Cited on p. 111.)

Mayers, D. F. 1966. Convergence of polynomial interpolation. Chap. 3, pages 15–26 of: Handscomb, D. C. (ed), *in Methods of Numerical Approximation, Chapter 3*. Oxford: Pergamon Press. (Cited on p. 158.)

McLachlan, R. I., and Quispel, G. R. W. 2002. Splitting methods. *Acta Numerica*, **11**, 341–434. (Cited on p. 68.)

Micchelli, C. A. 1986. Interpolation of scattered data: Distance matrices and conditionally positive definite functions. *Constr. Approx.*, **2**, 11–22. (Cited on p. 93.)

Mickens, R. E. 2002. Nonstandard finite difference schemes for differential equations. *J. Difference Eq. and Applic.*, **8**(9), 823–847. (Cited on p. 194.)

Mickens, R. E. 2020. *Nonstandard Finite Difference Schemes: Methodology and Applications*. World Scientific. (Cited on p. 194.)

Mills, S. 1985. The independent derivation of Leonhard Euler and Colin MacLaurin of the Euler-MacLaurin summation formula. *Arch. Hist. Exa. Sci.*, **33**, 1–13. (Cited on p. 126.)

Mishra, P. K., Nath, S. K., Sen, M. K., and Fasshauer, G. E. 2018. Hybrid Gaussian-cubic radial basis functions for scattered data interpolation. *Comput. Geosci.*, **22**, 1203–1218. (Cited on p. 95.)

Mittal, R., and Iaccarino, G. 2005. Immersed boundary methods. *Ann. Rev. Fluid Mech.*, **37**, 239–261. (Cited on p. 83.)

Mokhtari, R., and Mostajeran, F. 2020. A high order formula to approximate the Caputo fractional derivative. *Comm. Appl. Math. Comp.*, **2**, 1–29. (Cited on p. 144.)

Morton, K. W., and Mayers, D. F. 1994. *Numerical Solution of Partial Differential Equations*. Cambridge University Press. (Cited on pp. 49, 51, and 83.)

Nakatsukasa, Y., Sète, O., and Trefethen, L. N. 2018. The AAA algorithm for rational approximation. *SIAM J. Sci. Comput.*, **40**(3), A1494–A1522. (Cited on p. 156.)

Nehari, Z. 1995 (originally published by Mc-Graw Hill 1952). *Conformal Mapping*. Dover. (Cited on p. 123.)

Neidinger, R. D. 2010. Introduction to automatic differentiation and MATLAB object-oriented programming. *SIAM Rev.*, **52**(3), 545–563. (Cited on p. 31.)

Nielsen, M. E., and Fornberg, B. 2024. High-order numerical method for solving elliptic partial differential equations on unfitted node sets. ArXiv 2407.15825. (Cited on pp. 110 and 208.)

Nowak, R. 2019. Convergence acceleration of alternating series. *Num. Algor.*, **81**, 591–608. (Cited on p. 138.)

Olver, F. W. J., Lozier, D. W., Boisvert, R. F., and Clark, C. W. 2010. *NIST Handbook of Mathematical Functions*. Cambridge University Press. (Cited on pp. 33 and 144.)

Ong, B. W., and Spiteri, R. J. 2020. Deferred correction methods for ordinary differential equations. *J. Sci. Comput.*, **83**, Article Nr: 60. (Cited on p. 45.)

Oskooi, A., and Johnson, S. G. 2011. Distinguishing correct from incorrect PML proposals and a corrected unsplit PML for anisotropic, dispersive media. *J. Comput. Phys.*, **230**(7), 2369–2377. (Cited on p. 84.)

Ossendrijver, M. 2016. Ancient Babylonian astronomers calculated Jupiter's position from the area under a time-velocity graph. *Science*, **351**, 482–484. (Cited on p. 124.)

Pareschi, L., and Russo, G. 2005. Implicit-explicit Runge-Kutta schemes and applications to hyperbolic systems with relaxation. *J. Sci. Comput.*, **25**(1), 129–155. (Cited on p. 67.)

Pathria, D. 1997. The correct formulation of intermediate boundary conditions for Runge-Kutta time integration of initial boundary value problems. *SIAM J. Sci. Comput.*, **18**(5), 1255–1266. (Cited on p. 50.)

Pearson, J. 2009. *Computation of hypergeometric functions*. M.Phil. thesis, University of Oxford. (Cited on p. 33.)

Pearson, J. W., Olver, S., and Porter, M. A. 2017. Numerical methods for the computation of the confluent and Gauss hypergeometric functions. *Num. Algor.*, **74**, 821–866. (Cited on p. 33.)

Pereyra, V. 1966. On improving an approximate solution of a functional equation by deferred corrections. *Numerische Mathematik*, **8**, 376–391. (Cited on p. 45.)

Persson, P.-O., and Strang, G. 2004. A simple mesh generator in MATLAB. *SIAM Rev.*, **46**(2), 329–345. (Cited on pp. 109 and 207.)

Peskin, C. S. 1977. Numerical analysis of blood flow in the heart. *J. Comput. Phys.*, **25**, 220–252. (Cited on p. 83.)

Peskin, C. S. 2002. The immersed boundary method. *Acta Numerica*, **11**, 479–517. (Cited on p. 83.)

Peters, G., and Wilkinson, J. H. 1979. Inverse iteration, ill-conditioned equations and Newton's method. *SIAM Rev.*, **21**(3), 339–360. (Cited on p. 82.)

Petras, A., Ling, L., Piret, C., and Ruth, S. J. 2019. A least-squares implicit RBF-FD closest point method and applications to PDEs on moving surfaces. *J. Comput. Phys.*, **381**, 146–161. (Cited on p. 111.)

Pinelis, I. 2018. An alternative to the Euler-Maclaurin summation formula: Approximating sums by integrals only. *Numerische Mathematik*, **140**, 755–790. (Cited on p. 131.)

Piret, C. 2012. The orthogonal gradients method: A radial basis functions method for solving partial differential equations on arbitrary surfaces. *J. Comput. Phys.*, **231**, 4662–4675.

Platte, R. B., Trefethen, L. N., and Kuijlaars, A. B. J. 2011. Impossibility of fast stable approximation of analytic functions from equispaced samples. *SIAM Rev.*, **53**(2), 308–318. (Cited on p. 14.)

Podlubny, I. 1990. *Fractional Differential Equations*. Academic Press. (Cited on p. 142.)

Pooladi, F., and Larsson, E. 2024. Stabilized interpolation using radial basis functions augmented with select radial polynomials. *J. Comp. Appl. Math.*, **437**, Article Nr: 115482. (Cited on p. 95.)

Powell, M. J. D. 1981. *Approximation Theory and Methods*. Cambridge University Press. (Cited on pp. 159 and 164.)

Powell, M. J. D. 1992. *The Theory of Radial Basis Functions Approximation in 1990*. Clarendon Press. Chap. 3, Advances in Numerical Analysis II: Wavelets, Subdivision Algorithms, and Radial Basis Functions, ed. Will Light, pages 105–210. (Cited on p. 92.)

Powell, M. J. D. 2005. *Five lectures on radial basis functions*. Tech. rept. IMM-2005-03. Technical University of Denmark, DTU, Lyngby. (Cited on p. 93.)

Prautzsch, H., Boehm, W., and Paluszny, M. 2002. *Bézier and B-Spline Techniques*. Springer. (Cited on p. 159.)

Ralston, A., and Rabinowitz, P. 1978. *A First Course in Numerical Analysis*. International series in pure and applied mathematics. McGraw-Hill. (Cited on p. 126.)

Ramezani, M., Mokhtari, R., and Haase, G. 2020. Some high order formulae for approximating Caputo fractional derivatives. *Appl. Numeric. Math.*, **153**, 300–318. (Cited on p. 144.)

Reddy, S. C., and Trefethen, L. N. 1992. Stability of the method of lines. *Numerische Mathematik*, **62**, 235–267. (Cited on p. 50.)

Reeger, J. A., and Fornberg, B. 2018. Numerical quadrature over smooth surfaces with boundaries. *J. Comput. Phys.*, **355**, 176–190. (Cited on p. 200.)

Renou, M-O., Acin, A., and Navascués, M. 2023. Quantum physics falls apart without imaginary numbers. *Sci. Am.*, **328**(4), 62–67. (Cited on p. 112.)

Richardson, L. F. 1911. The approximate arithmetical solution by finite differences of physical problems involving differential equations, with an application to a masonry dam. *Philos. Trans. R. Soc. London*, **210**, 307–357. (Cited on p. 1.)

Rosenbrock, H. H. 1960. Some general implicit processes for the numerical solution of differential equations. *Comput. J.*, **5**(4), 329–330. (Cited on p. 67.)

Runge, C. 1901. Über empirische Functionen und die Interpolation zwischen äquidistanten Ordinaten. *Z. Math. Phys.*, **46**, 224–243. (Cited on p. 13.)

Ruprecht, D. 2018. Wave propagation characteristics of Parareal. *Comput. Vis. Sci.*, **19**(1–2), 1–17. (Cited on p. 200.)

Rutishauser, H. 1976. *Vorlesungen über numerische Mathematik*. Vol. 1., English translation, Lectures on Numerical Mathematics, W. Gautschi, ed., Birkhäuser, Boston, 1990: Birkhäuser. (Cited on p. 156.)

Ruuth, S. J., and Merriman, B. 2008. A simple embedding method for solving partial differential equations on surfaces. *J. Comput. Phys.*, **227**(3), 1943–1961. (Cited on p. 111.)

Sagaut, P., Suman, V. K., Sundaram, P., Rajpoot, M. K., Bhumkar, Y. G., Sengupta, S., Sengupta, A., and Sengupta, T. K. 2023. Global spectral analysis: Review of numerical methods. *Comp. Fluids*, **261**, Article Nr: 105915. (Cited on p. 195.)

Samarin, E. N., and Chudov, L. A. 1963. On the stability of the numerical integration of systems of ordinary differential equations arising in the use of the straight line method. *USSR Computational Math. and Math. Physics*, **3**(6), 1537–1543. (Cited on p. 50.)

Samko, S. G., Kilbas, A. A., and Marichev, O. I. 1993. *Fractional Integrals and Derivatives: Theory and Applications*. Gordon and Breach Science Publishers. (Cited on p. 142.)

Santos, L. G. C., Manzanares-Filho, N., Menon, G. J., and Abreu, E. 2018. Comparing RBF-FD approximations based on stabilized Gaussians and on polyharmonic splines with polynomials. *Int. J. Numer. Methods Eng.*, **115**, 462–500. (Cited on p. 104.)

Schaback, R. 1995. Error estimates and condition numbers for radial basis function interpolants. *Adv. Comput. Math.*, **3**, 251–264. (Cited on p. 94.)

Schiesser, W. E. 1991. *The Numerical Method of Lines*. Academic Press. (Cited on p. 50.)

Schoenberg, I. J. 1938. Metric spaces and completely monotone functions. *Ann. Math.*, **39**, 811–841. (Cited on p. 93.)

Schoenberg, I. J. 1946. Contributions to the problem of approximation of equidistant data by analytic functions. *Quart. Appl. Math.*, **4**, 45–99 and 112–141. (Cited on p. 159.)

Schumaker, L. L. 1981. *Spline Functions: Basic Theory*. John Wiley & Sons. (Cited on p. 159.)

Scott, J., and Tůma, M. 2022. Solving large linear least squares problems with linear equality constraints. *BIT*, **62**, 1765–1787. (Cited on p. 183.)

Shankar, V., and Fogelson, A. L. 2018. Hyperviscosity-based stabilization for radial basis function-finite difference (RBF-FD) discretizations of advection-diffusion equations. *J. Comput. Phys.*, **372**, 616–639. (Cited on p. 106.)

Shankar, V., Wright, G. B., and Fogelson, A. L. 2021. An efficient high-order meshless method for advection-diffusion equations on time-varying irregular domains. *J. Comput. Phys.*, **445**, Article Nr: 110633. (Cited on p. 110.)

Shapira, Y. 2008. *Matrix-based Multugrid: Theory and Applications*. Springer. (Cited on p. 83.)

Sidi, A. 2003. *Practical Extrapolation Methods*. Cambridge Monographs on Applied and Computational Mathematics, vol. 10. Cambridge University Press. (Cited on pp. 138 and 185.)

Slak, J., and Kosec, G. 2019. On generation of node distributions for meshless PDE discretizations. *SIAM J. Sci. Comput.*, **41**(5), A3202–A3229. (Cited on p. 207.)

Smith, G. D. 1965. *Numerical Solution of Partial Differential Equations*. Oxford Mathematical Handbooks. Oxford University Press. (Cited on p. 111.)

Smolyak, S. A. 1963. Quadrature and interpolation formulas for tensor products of certain classes of functions. *Dokl. Akad. Nauk SSSR*, **4**, 240–243. (Cited on p. 205.)

Southwell, R. V. 1945. On relaxation methods: A mathematics for engineering science. *Proc. R.Soc. London, Ser. A*, **184**(998), 253–288. (Cited on p. 205.)

Spotz, W. F., and Carey, G. F. 1995. High-order compact scheme for the stream-function vorticity equations. *Int. J. Numer. Meth. Eng.*, **38**, 3497–3512. (Cited on p. 71.)

Spotz, W. F., and Carey, G. F. 2001. Extension of high-order compact schemes to time-dependent problems. *Numer. Meth. for Part. Diff. Eq.*, **17**(6), 657–672. (Cited on p. 71.)

Squire, W., and Trapp, G. 1998. Using complex variables to estimate derivatives of real functions. *SIAM Rev.*, **40**, 110–112. (Cited on p. 114.)

Stahl, H. 1997. The convergence of Padé approximants to functions with branch points. *J. Approx. The.*, **91**(2), 139–204. (Cited on p. 191.)

Strang, G. 1968. On construction and comparison of difference schemes. *SIAM J. Numer. Anal.*, **5**, 506–516. (Cited on p. 68.)

Strikwerda, J. C. 2004. *Finite Difference Schemes and Partial Differential Equations*. 2nd ed. SIAM. (Cited on p. 49.)

Stüben, K. 2002. A review of algebraic multigrid. *J. Comp. Appl. Math.*, **128**(1–2), 281–309. (Cited on p. 83.)

Süli, E., and Mayers, D. 2003. *An Introduction to Numerical Analysis.* Cambridge University Press. (Cited on p. 42.)

Sutmann, G., and Steffen, B. 2006. High-order solvers for the three-dimensional Poisson equation. *J. Comp. Appl. Math.*, **187**, 142–170. (Cited on p. 74.)

Svärd, M., and Nordström, J. 2014. Review of summation-by-parts schemes for initial-boundary-value problems. *J. Comput. Phys.*, **268**, 17–38. (Cited on p. 57.)

Taflove, A., Hagness, S. C., and Piket-May, M. 2005. Computational electromagnetics: The finite-difference time-domain method. ch 9 Pages 629–670 of: Chen, W.-K. (ed.), *The Electrical Engineering Handbook.* Elsevier. (Cited on p. 61.)

Taha, T. R., and Ablowitz, M. J. 1984. Analytical and numerical aspects of certain nonlinear evolution equations. II. Nonlinear Schrödinger equation. *J. Comput. Phys.*, **55**, 203–230. (Cited on p. 68.)

Tarwater, A. E. 1985. *Parameter study of Hardy's multiquadric method for scattered data interpolation.* Technical Report UCRL-54670. Lawrence Livermore National Laboratory, Livermore, CA. (Cited on p. 93.)

Thompson, J. F., Soni, B. K., and Weatherill, N. P. (eds.). 1998. *Handbook of Grid Generation.* CRC Press. (Cited on p. 123.)

Tolstykh, A. I. 2000. Using RBF-based differencing formulas for unstructured and mixed structured-unstructured grid calculations. Pages 4606–4624 of: *Proceedings of the 16th IMACS World Congress*, vol. 228. Conference held in Lausanne 2000, Proceedings Editors: M. Deville and R. Owens. (Cited on p. 101.)

Toro, E. F. 2009. *Riemann Solvers and Numerical Methods for Fluid Dynamics: A Practical Introduction.* Springer. (Cited on p. 71.)

Trefethen, L. N. 2000. *Spectral Methods in MATLAB.* SIAM. (Cited on pp. 13 and 17.)

Trefethen, L. N. 2013, extended edition 2019. *Approximation Theory and Approximation Practice.* SIAM. (Cited on pp. 6, 13, 17, and 158.)

Trefethen, L. N. 2020. Quantifying the ill-conditioning of analytic continuation. *BIT*, **60**, 901–915. (Cited on p. 190.)

Trefethen, L. N. 2023. Numerical analytic continuation. *JPN J. Ind. Appl. Math.*, **40**, 1587–1636. (Cited on pp. 35 and 190.)

Trefethen, L. N., and Weideman, J. A. C. 2014. The exponentially convergent trapezoidal rule. *SIAM Rev.*, **56**, 384–458. (Cited on p. 126.)

Trefethen, L. N., Birkisson, Á., and Driscoll, T. A. 2017. *Exploring ODEs.* SIAM. (Cited on p. 26.)

Turnbull, H. W. (ed.). 1939. *James Gregory. Tricentenary Memorial Volume.* G. Bell & Sons Ltd. (Cited on p. 128.)

Valério, D., Ortigueira, M. D., and Lopes, A. M. 2022. How many fractional derivatives are there? *Mathematics*, **10**, Article Nr: 737. (Cited on p. 141.)

van der Sande, K., and Fornberg, B. 2021. Fast variable density 3-D node generation. *SIAM J. Sci. Comput.*, **43**(1), A242–A257. (Cited on pp. 207 and 210.)

Verwer, J. G. 2007. On time staggering for wave equations. *J. Sci. Comput.*, **33**, 139–154. (Cited on p. 58.)

Vichnevetsky, R. 1987. Wave-propagation analysis of difference-schemes for hyperbolic equations – A review. *Int. J. for Num. Meth. Fluids*, **7**(5), 409–452. (Cited on p. 195.)

Wegert, E. 2012. *Visual Complex Functions.* Birkhäuser–Springer. (Cited on p. 112.)

Wegert, E. 2024. About the cover: Complex finite differences of higher order. *Comp. Meth. Funct. Ther.*, **24**(1), 1–6. (Cited on pp. 120 and 121.)

Weideman, J. A. C. 2002. Numerical integration of periodic functions: A few examples. *Amer. Math. Monthly*, **109**(1), 21–36. (Cited on p. 126.)

Wendland, H. 2005. *Scattered Data Approximation*. Cambridge Monographs on Applied and Computational Mathematics, vol. 17. Cambridge University Press. (Cited on p. 89.)

Wendland, H., and Künemund, J. 2020. Solving partial differential equations on (evolving) surfaces with radial basis functions. *Adv. Comput. Math.*, **46**, Article Nr: 64. (Cited on p. 111.)

Wilbraham, H. 1848. On a certain periodic function. *Cambridge & Dublin Math. J.*, **3**, 198–201. (Cited on p. 22.)

Willers, I. M. 1974. A new integration algorithm for ordinary differential equations based on continued fraction approximations. *Comm. ACM*, **17**, 504–508. (Cited on p. 30.)

Winograd, S. 1978. On computing the discrete Fourier transform. *Math. Comput.*, **32**, 175–199. (Cited on p. 179.)

Womersley, R. 2020. *Interpolation and cubature on the sphere*. https://web.maths.unsw.edu.au/~rsw/Sphere/ (Cited on pp. 105 and 107.)

Wright, G. B., and Fornberg, B. 2006. Scattered node compact finite difference-type formulas generated from radial basis functions. *J. Comput. Phys.*, **212**, 99–123. (Cited on p. 109.)

Wright, G. B., and Fornberg, B. 2017. Stable computations with flat radial basis functions using vector-valued rational approximations. *J. Comput. Phys.*, **331**, 137–156. (Cited on pp. 95 and 101.)

Wright, G. B., Flyer, N., and Yuen, D. A. 2010. A hybrid radial basis function-pseudospectral method for thermal convection in a 3D shperical shell. *Geophys., Geochem., Geosyst.*, **11**, Q07003. (Cited on pp. 97 and 98.)

Wright, G. B., Jones, A., and Shankar, V. 2023. MGM: A meshfree geometric multilevel method for systems arising from elliptic equations on point cloud surfaces. *SIAM J. Sci. Comput.*, **45**(2), A312–A337. (Cited on pp. 110 and 111.)

Yang, J. 2010. *Nonlinear Waves in Integrable and Nonintegrable Systems*. SIAM. (Cited on p. 66.)

Yee, K. S. 1966. Numerical solution of initial value problems involving Maxwells equations in isotropic media. *IEEE Trans. Ant. Propog.*, **14**, 302–307. (Cited on pp. 58 and 61.)

Yoshida, H. 1990. Construction of higher order symplectic integrators. *Phys. Lett. A*, **150**, 262–268. (Cited on p. 68.)

Yu, P. X., and Tian, Z. F. 2019. A high-order compact scheme for the pure streamfunction (vector potential) formulation of the 3D steady incompressible Navier-Stokes equations. *J. Comput. Phys.*, **382**, 65–85. (Cited on p. 71.)

Yuksel, C. 2015. Sample elimination for generating Poisson disk sample sets. *Comp. Graph. Forum*, **34**, 25–35. (Cited on p. 210.)

Zabusky, N. J., and Kruskal, M. D. 1965. Interactions of "solitons" in a collisionless plasma and the recurrence of initial states. *Phys. Rev. Lett.*, **15**, 240–243. (Cited on p. 65.)

Zamolo, R., Nobile, E., and Šarler, B. 2019. Novel multilevel techniques for convergence acceleration in the solution of systems of equations arising from RBF-FD meshless disretizations. *J. Comput. Phys.*, **392**, 311–334. (Cited on p. 111.)

Zayernouri, M., Wang, L. L., Shen, J., and Karniadakis, G. E. 2024. *Spectral and Spectral Element Methods for Fractional Ordinary and Partial Differential Equations*. Cambridge Monographs on Applied and Computational Mathematics, vol. 41. Cambridge University Press. (Cited on p. 151.)

Zheng, Z., and Li, X. 2022. Theoretical analysis of the generalized finite difference method. *Comp. Math. Applic.*, **120**, 1–14. (Cited on p. 85.)

Zhuang, Q., Heryudono, A., Zeng, F., and Zhang, Z. 2024. Collocation methods for integral fractional Laplacian and fractional PDEs based on radial basis functions. *Applied Mathematics and Computation*, **469**, Article Nr: 128548. (Cited on pp. 151 and 154.)

Zingg, D. W. 2000. Comparison of high-accuracy finite-difference methods for linear wave propagation. *SIAM J. Sci. Comput.*, **22**(2), 476–502. (Cited on p. 195.)

Zingg, D. W., and Lomax, H. 1993. Finite-difference schemes on regular triangular grids. *J. Comput. Phys.*, **108**(2), 306–313. (Cited on p. 205.)

Zlatev, Z., Dimov, I., Faragó, I., and Havasi, Á. 2018. *Richardson Extrapolation: Practical Aspects and Applications*. De Gruyter Series in Applied and Numerical Mathematics, vol. 2. De Gruyter. (Cited on p. 186.)

Index

Printed by Integrated Books International,
United States of America